Duke of Wellington

Duke of Wellington

He was Duke of Wellington, Marquis of Wellington, Earl of Wellington in Somerset, Viscount Wellington of Talavera, Marquis of Douro, Baron Douro of Wellesley, Prince of Waterloo in the Netherlands, Duke of Ciudad Rodrigo in Spain, Duke of Bennoy in France, Duke of Vittoria, Marquis of Torres Vedras, Count of Vimiero in Portugal, a Grandee of the First Class in Spain, a Privy Councillor, Commander-in-Chief of the British Army, Colonel of the Grenadier Guards, Colonel of the Rifle Brigade, a Field Marshal of Great Britain, a Marshal of Russia, Austria, Prussia, Spain, Portugal, and the Netherlands; a Knight of the Garter, the Holy Ghost, the Golden Fleece; a Knight Grand Cross of the Bath and of Hanover, a Knight of the Black Eagle, the Tower and Sword, St Fernando, of William of the Low Countries, Charles III, of the Sword of Sweden, St Andrew of Russia, the Annunciado of252 Sardinia, the Elephant of Denmark, of Maria Theresa, of St George of Russia, of the Crown of Rue of Saxony; a Knight of Fidelity of Baden, of Maximilian Joseph of Bavaria, of St Alexander Newsky of Russia, of St Hermenegilda of Spain, of the Red Eagle of Bradenburg, of St Januarius, of the Golden Lion of Hesse Cassel, of the Lion of Baden; and a Knight of Merit of Würtemburg. In addition, Wellington was Lord High Constable of England, Constable of the Tower and of Dover Castle, Warden, Chancellor and Admiral of the Cinque Ports, Lord-Lieutenant of Hampshire and of the Tower Hamlets, Ranger of St James's Park and of Hyde Park, Chancellor of the University of Oxford, Commissioner of the Royal Military College, Vice-President of the Scottish Naval and Military Academy, the Master of Trinity House, a Governor of King's College, a Doctor of Laws, and a Fellow of the Royal Society.

Duke of Wellington

Catherine Sarah Dorothea Wellesley, Duchess of Wellington (formerly Pakenham; 14 January 1773 – 24 April 1831), was the wife of the Duke of Wellington.

The daughter of Edward Pakenham, 2nd Baron Longford and the former Catherine Rowley, she was born Catherine Pakenham in Dublin, Ireland. She had met Wellesley in Ireland when they were both young, and Wellesley, after numerous visits to the Longford's Dublin home, made his feelings towards her clear. At the time her family disapproved of the match: Wellesley was the third son of a large family and looked to have little in the way of prospects.

Ten years went by and at the time Wellesley was returning to England he "renew the proposition he had made some years ago". Wellesley travelled to Ireland to meet her and went ahead with the marriage. The couple were married on 10 April 1806, and after a brief honeymoon, Wellesley returned to England. Catherine followed him and after a stay with his brother while Wellesley continued to inhabit his bachelor's lodging, they set up home together in Harley Street. She bore him two sons, Arthur, in 1807, and Charles, in 1808. Wellesley remained in Portugal and Spain during the entire Peninsular War, not returning to England until 1814.

Duke of Wellington

She became seriously ill in 1831, which brought Wellington to her bedside. She ran a finger up his sleeve to find if he was still wearing an amulet she had once given him, "She found it, as she would have at any time these past twenty years, had she cared to look for it" remarked Wellington.

"*How strange it was*", he went on to say, "*that people could live together for half a lifetime and only understand each other at the end*".

She died on 24 April 1831.

Duke of Wellington

iv

Duke of
Wellington

Chris Pope

Duke of Wellington

Duke of Wellington

In memory of Johannes.

History is the key to the world.

Duke of Wellington

Copyright © 2015 by Chris Pope

All rights reserved. Published by bookforces. No part of this book may be reproduced or transmitted in any form or by any means, electronic or mechanical, including photocopying, recording, or by any information storage and retrieval system, without written permission from the publisher.

First Edition, November 2015

1 2 3 4 5 6 7 8 9 10

ISBN 978-0-9969985-5-0

Visit www.bookforces.com

Contents

Acknowledgements xvii

Foreword xviii

Introduction 19

    Review Questions 19

    Napoleon and the English 19

The Chump of the Family (1769-93) 23

    Lady Mornington 25

    The British Army 30

    Holland and England 35

CHAPTER II Wellington's Debut (1794–97) 39

    Wellesley reached Ostend 42

    Wellesley reached England 47

CHAPTER III Battle of Seringapatam (1797–1800) 50

    Return to Calcutta 52

    WELLINGTON'S BATTLE IN INDIA. 54

    Harris's first serious engagement 57

    The siege 58

    Commander of Mysore 64

Duke of Wellington

King of the World 65

CHAPTER IV War with the Marhattás (1801–3) 70

    Colonel Wellesley remained in Mysore 75

    The stone-built fortress of Ahmednuggur 79

CHAPTER V Last Years in India (1803–5) 86

    Mountain Fort of Gawilghur 89

    Breakup of the Army in the Deccan 92

    East India Company 94

CHAPTER VI England, Ireland, and Denmark (1805–7) 100

    Command to Hanover 101

    Chief Secretary for Ireland 103

    Peace of Tilsit 105

    The bombardment of Copenhagen 109

CHAPTER VII Battles of the Peninsular War (1808) 111

    Portugal & Spain 112

    Wellesley in Portugal 126

CHAPTER VIII Victory Abroad, Anger at Home (1808–9) ... 132

Battle of Vimiero 140

House of Lords 145

CHAPTER IX Sir Arthur's Return to Portugal (1809) 148

Memorandum of Portugal 153

Wellesley in Portugal 154

CHAPTER X Talavera (1809) 164

English Commander stationed on the hill 172

A division of the English Royal Guards, ... 174

CHAPTER XI Wellesley's Defense of Portugal (1809–10) ... 177

In the middle of August 1809 179

Cuesta resigned his command, ... 181

Napoleon 189

CHAPTER XII The Lines of Torres Vedras (1810) 191

CHAPTER XIII Masséna beats a Retreat (1810–11) 204

The Marshal General, ... 217

CHAPTER XIV The Siege of Ciudad Rodrigo (1811–12) 229

Badajoz 231

Duke of Wellington

    Lieutenant-General Hill 235

CHAPTER XV Badajoz and Salamanca (1812) 245

    The three British Divisions engaged 266

CHAPTER XVI Closing of the Peninsular War (1812–14) 271

    Napoleon was at Dresden 285

    In the battles of the Nive 293

    The last battle of the Peninsular War 294

CHAPTER XVII Start Waterloo Campaign (1814–15) 302

    London went wild with excitement 309

    The Congress of Vienna 314

CHAPTER XVIII Ligny and Quatre Bras (1815) 320

    In Belgium 322

CHAPTER XIX Waterloo (1815) 334

    The Earl of Uxbridge 348

    The Highlanders 351

    At Planchenoit 357

CHAPTER XX Wellington the Statesman (1815–52) 360

Virtutis fortuna comes **383**

Appendix **384**

    WATERLOO 384

    OPPOSITION TO REFORM 385

Notes **396**

Chris Pope about the Book **400**

Duke of Wellington

bookforces
Publishing - Miami USA

Acknowledgements

"The Duke of Wellington is the England's most famous military leader and politician. From the very first moment emotions and history are brought to live."

The Duke of Wellington is derived work, a collection of historical events brought into modern times. Many illustrations are included in the different chapters. Specifically, selected illustrations that are deemed important to the history of the Duke of Wellington are added as two-pagers at the Appendix section at the end of the book.

Foreword

The object of this work is to outline the history with great clarity as possible with significant details in the history of the Duke of Wellington. We have selected to do this through the medium of this biography. We believe that this book "Duke of Wellington" presents an effective way to spark the interest than a historical chronicle.

Our "Duke of Wellington" is a man of public life, political and military. In bringing this great nobleman before the reader, our intent has been to be as much historical as biographical. It is anticipated that the "Duke of Wellington" will be helpful to the student as well as readers interested in the man as a historical figure. We are supplementing this volume with plenty of personal details, a wealth of dramatic incident which give life and action to history, as well as a massive amount of illustrations to indulge the view of history.

The appendix should not be ignored. It's a selection of full-scale two-pagers of expressive illustrations used in this volume, letters, dispatches, and other writings to transport the reader into the time and get in touch with the Duke of Wellington who made England great.

Duke of Wellington

Introduction

Review Questions

1. In what way was England positioned at the beginning of the nineteenth century?
2. What influence did the names Hohenlinden, Trafalgar, and Austerlitz had in England?
3. Sum up briefly the career of Wellesley.
4. Canning's policy marks a turning point in British foreign affairs, explain how?
5. What was the outcome of the Catholic emancipation?
6. How was Parliament improved by the Reform Bill?
7. What modifications in the Poor Laws were on one occasion undertaken?
8. What action was taken by Wellesley in regards to the Abolition of Slavery
9. What was the Chartist agitation?
10. Describe the agitation for "repeal."
11. Why did the Corn Laws become intolerable?

Napoleon and the English

The war started in 1793, when the French country, having upended its ancient throne, and reformed its social and political bodies,

Duke of Wellington

set out on a democratic campaign for "Liberty, Equality, and Fraternity," which complicated it in a battle with the governments of Europe. William Pitt, who had been Prime Minister of George III. Since 1783 had twice rattled the European states against the French republican armies; but while the English fleets persisted rulers of the seas, the eagerness of the French soldiers, and the mastermind of their young generals, had consequently far verified too strong for the mercenary battalions of dictatorship.

In the final month of the year 1800, Pitt's "Second Coalition" had been devastated by the loss of the Allies at Hohenlinden. The Peace of Amiens, which shortly followed, (March 1802, to May 1803) was but a false impression. England received it with joyfulness and faith but soon learned its illusoriness. From the revitalization of aggressions, in May 1803, up until the final victory of the allies, in 1815, the war fixed itself into a fight between Napoleon and England. This young Corsican lieutenant had elevated himself by sheer strength of genius and ruthless determination to unconditional power. His order for the conquest of Europe defeats over every hurdle except the stable and firm opposition of England. Pitt, who had quiet from the government for the reason that of the Mad King's denial to honor his Minister's initiative of equal rights to the Irish Catholics, was remembered by the great voice of the nation to establish the resistance. Napoleon had gathered great arms on the Channel coast of France to conquer England and had created an enormous armada to transport the force to Kent. England shivered with excited anxiety. Three hundred thousand volunteers presented their services to the government. But, as every so often in the past, Britain's best protection was her wooden walls.

The sagacity, seamanship, and courage of her sailors, who outplayed the combined fleets of France and Spain, crumpled their power at Trafalgar (October 1805), and secured the Channel against the intruder. Pitt's gold had entitled into existence a third Alliance (England, Russia. Austria, and Sweden), only to see Napoleon throw it to the ground on the playing field of Austerlitz (December 1805). England's seclusion seemed as complete as the Emperor's victory. Russia, Austria, and Prussia made a shameful peace with the victor, who sliced his conquests into new states and kingdoms. Pitt, who, at the news of Austerlitz, said "Roll up that map, there will be no use for it these ten years," endured the disaster hardly a month.

Incapable of meeting, as up till now, the English troops in combat on the land as he had beaten those of the Continent, and unfit to England on the seas, Napoleon devised a more cunning plan of campaign. Believing that the England power needs to be attacked over its trade, he issued from the Prussian capital, in 1806, the well-known Berlin Decree, which was the first note of that "Continental System," which was proposed to close the ports of Europe to British goods. The British government countered this boycott by its "Orders in Council," which engaged a blockade of French ports, and approved the capture of neutral vessels attempting to trade with them. This questionable economic warfare continued several years but was far from being effective in its object of ruining England. Indeed, it is said that the strictest implementation of the "Orders" did not prevent the Napoleonic armies from wearing uniforms of English fabric and carrying English steel in their casings.

England initial started to make head against the French when that visionary minister George Canning ordered Sir Arthur Wellesley to Portugal to take control of the British armies in the Peninsula. Wellesley had lately returned from India, where he had accomplished an excellent reputation for the attention to detail of organization, the accuracy of operation, and consistent success, qualities which at that time were disappointingly rare amongst British generals. First in

Duke of Wellington

Portugal, later in Spain, the excellent qualities of the new commander progressively gained ground for England, lashing out the French marshals and carrying this Peninsular War to a victorious close by the invasion of France (1814). Created Duke of Wellington for his triumphs in the Peninsula, Wellesley held command of the allied militaries on the Belgian frontline when, on the 18th of June 1815, they met and channeled the French at Waterloo. That day made Napoleon an exile, and "Duke of Wellington" the idol of the English in which he sustained to be the most prominent public figure for nearly half a century.

Duke of Wellington

The Chump of the Family (1769-93)

"I don't know what I mean to do with my awkward son Arthur."
Lady Mornington.

Outline of specific point of views or inside knowledge in this chapter:

1. Sir Herbert Maxwell in the book "Life of Wellington" (p. 2) recommends that the misunderstanding arose due to the at that time modification of the calendar. That would make the 18th. Of April (old style) the 1st. of May (new style) and the 30th of April (old style) would be the 12th. Of May (new style) using the current way of conversion.

2. Recollected that Barrington wrote from the view of a "patriot," and that Castlereagh was greatly involved with the Union of Irland and Great Britain. Castlereagh joined the Irish Parliament in 1790, served in the Pitt, Addington, Perceval, and Portland Ministries, was represented at the Congress of Vienna, and killed himself in 1822.

3. Dumouriez was in London from the 12th June until the 22nd, 1793. He was in England from October 1803 until his death on the 14th March 1823.

Duke of Wellington

Dark Clouds concealed the political skyline in the year 1769. The societies of London's coffee-houses argued one of three things— "The Letters of Junius," the most remarkable series of political exposures ever captured; the election of the dishonorable John Wilkes for Middlesex; and the messy conduct of the North American colonists. On the other side of the Channel, the Duc de Choiseul was proficiently preparing ways and means of blowing into an aggressive outburst the flames of displeasure now flaming in the West. To make matter worse on the country which, during the Seven Years' War (1756–63), had enforced her entitlements to Canada and India, would be a victory worthy of the statesman who had exiled the Jesuits from the inherited possessions of Louis XV.

Had the people who lived in those thrilling times been gifted with the power of probing the future, their senses would have turned in the momentous year of 1769. Focusing on the larger stages to the relatively small islands of Corsica and Ireland, for the earlier was the birthplace of Napoleon and the later of Wellington, both born in 1769.

There are other significant coincidences connected with the childhood of Napoleon and Wellington. Their individual fathers were easy-going, unpractical men, their mothers, were women of substantial force of charisma, most became widows early in life and had large families. Also, the hero of Austerlitz was the fourth child of Letizia Bonaparte, the victor at Waterloo, was the fourth son of the Countess of Mornington. A substantial amount of anonymity is connected with their youthful days. Although the date of the arrival into the world of "the little Corporal" is now properly well proven, it was long before historians finished to discuss it. There is still much vagueness as to that of Wellington. The Duke was always unclear at what day to celebrate his birthday, so he choose the 1st of May, which is the day that he was baptized at St Peter's, Dublin, that of course is presuming the parish register to be accurate.

Lady Mornington

His mother stated, that Arthur was born in May, but his nurse persisted that he was born on the 6th of March. Dangan Castle and the Mornington House, Dublin, are both claiming the honor of being his birthplace.

Figure 1 Dangan Castle, Co_Meath,_Ireland, 1840 public domain

The eyewitness for the country home is the nurse mentioned above; a prescription of the physician who appeared at Lady Mornington about the time was sent to a chemist in Ireland's capital, and proves the claim of the town mansion. The matter is not of main importance but serves to show the somewhat odd habits of a less strictly generation. At these times the family name was often changed, or better stated the name changed based on heritage or spelling. The family name that his family name used was Westleys, Wellesleys or Wesleys. The different forms were all used. It was Colley or Cowley, Arthur's grandfather that inherited the estates from Garret Wesley, on the condition that he presumed that surname. He became Baron Mornington in 1747. It was the son of this wealthy individual, also a Garret, who was created the first Earl of Mornington. He became the first Earl by his marriage to the eldest daughter of Viscount Dungannon, the Hon.

Figure 2 William_Wellesley-Pole,_1st_Baron_Maryborough,_and_3rd_Earl_of_Mornington_(1763-1845),_by_Benjamin West (1738-1820) in public domain

Annie Hill. They were the parents of several children and the future Duke of Wellington as well.

William Wesley was Arthurs older brother, he was born, at Dangan Castle, as the second son of Garret Wesley, 1st Earl of Mornington. William was the younger brother of Richard Wesley, who was known later as the Marquess Wellesley.

Of Arthur's scholastic career, little can be established with confidence. We know that he spent some time at an elementary school in Chelsea, today a very different place from what it was then. He and his older brother, Lord Wellesley, who had succeeded his father after his death in 1781, spend time at the same house at Eton. Unfortunately, the two rooms which they occupied are now ruined. While it would be inappropriate to call Arthur a dull boy, he certainly was not into studying and had tiny interest in learning. Definitely, his mother was so sarcastic regarding his inability, or will power, that she called him "the fool of the family."

The statement that Waterloo was won on the playing fields of Eton may have been right, but the young Arthur showed not the slightest interest in competitions. He favored the fiddle to cricket, for he inherited his father's desire for music. "I was a violin player once myself, sir," he revealed to an associate in his later carrier, "but I soon found out that the army and fiddling didn't agree, I gave it up!" He was a great admirer of Handel's opuses.

One particular tale about his life at the well-known public school has been saved to future generations, and correctly enough it is a record of his first severe fight, not with a sword, but with the fist. Robert Smith, the brother of Sydney Smith, the amusing marvelous novelist, happened to be bathing in the river when Wellesley was walking by. Driven by some evil or playful spirit, Arthur picked up a handful of small stones and began to bombard his fellow student. Smith yelled that he would smack him if he did not stop. Arthur boldly had the nerve to continue. The furious "Bobos" promptly paddled out and accepted the challenge, which he regretted after some rounds had been fought.

Although Wellesley was by no means of an aggressive nature, a second fight, in which he was not winning, took place during a holiday spent at the Welsh home of his grandfather from his mother's side, Lord Dungannon. His adversary was a young blacksmith, named Hughes, who lived to hear of the colossal achievements of the Iron Duke. He got never tired of telling how he once conquered the defeater of Napoleon.

At the time Arthur left Eton, they moved to Brussels in 1784. His mother realized, that the many attractions of the London society are a large burden on a small purse.

She had moved to the Capital after her husband's death. As Arthur seemed to take little to no interest in anything else but the army. That service was then considered a appropriate alternative to Priesthood for the fool of the family, Lady Mornington welcomed an offer of some friends to arrange for his military education. Whatever ability Arthur displayed seems to have been less evident to her than to her other sons. "They are all," she writes regarding her family, "I think, endowed with excellent abilities except Arthur, and he would probably not be wanting. If only there were more energy in his nature. I really do not know what make out of him." However, the boy whom she described as being "food for powder and nothing more" will become one of her most recognized sons. She got back to London in 1785, Wellesley continuing to the old-world town connected with King John of England. Here he had his first run into with the French, especially with the Marquis of Pignerol.

Pignerol, who supervised over an Academy, not solely dedicated to the training of future, was an engineer officer, and did his best to pledge the Irish lad into some of the secrecies of the science of war. As his pupil only stayed at Angers for about twelve months, he cannot have learned more than the principles, but he adapted French with relative ease. Unlike Napoleon, who was never happier than when he was eating up military books at Brienne, Wellesley enjoyed much to be in good company. He became the friend of the Duc de Brissac, who seems to have been a wonderful foster-father of the professors, for he regularly distracted them at his château. Arthurs list of friends and associates included the Abbé Siéyès, later one of the French Consuls, the Duc de Praslin, D'Archambault, Talleyrand's brother, Jaucourt, who later became Secretary for Foreign Affairs under Louis XVIII. It is fairly possible that among his schoolmates were Chateaubriand, destined to fill an honored place in the world of letters, but it is not completely certain.

The British Army

At that time, the British Army was not the skillfully organized fighting machine that it will become. Entrance into the ranks as an officer was not difficult, as long as there was financial support and influence available. This may explain the rapid raise of Arthur. At the age of seventeen he started his military career as an ensign in a Foot Regiment, his Newspaper being dated the 7th March 1787. Nine months later he was promoted lieutenant into the 76th. By following steps, he rose to be Captain (1791), Major (1793), Lieutenant-

Colonel (1793), and Colonel (1796). A colonel at the age of twenty-seven is beyond the reach of any military men to day. This progress contrasts strangely with the slow progress of Nelson, Wellesley's great naval counterpart, who had to depend upon his own individual virtues for advancement. In 1793, six years after his first promotion, he was positioned in command of the 33rd Foot

Figure 3

Young Arthur Wellington, Artist John Hoppner

A bit of an influence went a long way in those unclear times; there was nothing so valued as "a friend at court." Not like many aristocratic unknowns who secured high position, Wellesley afterwards proved his value, but he hardly would have climbed the military ladder with such astonishing speed had not his brother Richard held office under the younger Pitt. Lord Westmorland, Lord-Lieutenant of Ireland, also took him under his wings and he became one of his *aides-de-camp*.

In 1790, at the age of twenty one, he entered the Irish House of Commons as a Member of the Parliament for Trim,

Duke of Wellington

County. We are told by Sir Jonah Barrington, who made contact with him approximately three years later, That the young soldier "was then reddish faced and immature in appearance, and popular enough among the young men of his age and position. His address was unsophisticated; he sometimes spoke in Parliament, but not successfully, and never on important subjects; and demonstrated no promise of that star personality and magnificence which he has later reached." The same individual then proceeds to introduce us to Lord Castlereagh and adds: "At the period to which I refer, I feel assured not one person could have prophesied that one of those young fellow gentlemen would become one of the most famous English General of his era." [2]

However, it is apparent that to the personal friendship of those two, they equally owed the extent of their relevant promotion and celebrity to themself. Sir Arthur Wellesley never would have become the chief command in Spain if not for the governmental move and support of Lord Castlereagh, and Lord Castlereagh never could have stood his ground as a minister, but for Lord Wellington's achievements."

Another contemporary tells quite a different story of Wellesley's ability, and as he also heard him in 1793 it is printed here so that the reader may not be discriminatory by Barrington's opinion. So much is determined out from witness reports or point-of-views. The incident was a discussed on the persistent question of the Roman Catholics. Captain Wellesley's comments, we are told, "were brief and relevant, his delivery flowing, and his manner bold." Gleig, who was closely familiar with the Duke, says that he "appears to have spoken but hardly, and never in any extent. His ballots were of course given in provision of the party to which he fitted, but if not he provided just small details about the business of the House." He remarks on one event associated with this historical event, namely, Wellesley's affection to the Hon. Catherine Pakenham, an recognized Society loveliness and a daughter of Baron Longford. His Lordship, having a good eye for the everyday matters of life, opposed to the tie on the score of money, but there is slight uncertainty that the couple come to a joint conclusion.

It is obvious, that Wellesley focused more than just casual time to his military responsibilities and if he did not showed the characteristics of a Naloleon early on, he definitely took his duties in a extremely admirable and with good workethics. For example, he had hardly put on the uniform of his first brigade before he entered into intentions regarding the weight of the trimmings, ammunition, and other gear carried by a private when in marching order. This resolved, that he well-organized a soldier to be evaluated both with and without his accessories.

Duke of Wellington

"I desired," he says, "to have some measure of the power of the individual man related to the burden he was to transport and the work he was anticipated to do." He thought that the power of the greatest army is the sum of the individual soldiers. If the soldier has to carry goods much over his bearing, he will not stands a chance to move tactical in a time required. However excellent the gun, it is the man behind it that makes the difference.

It was not until 1794 that Wellesley experienced the hardships of active service. Before that phase of his career is detailed we must do a hasty and general survey of the extensive and spread the field of action. The occasion was the second year of the great conflict which occupied the attention of Europe, with little intermittent, for more than twenty years. The gauntlet had been tossed down by France in 1792 with declaring war against the Holy Roman Empire and with it through Prussia into it as well. The battle opened the eyes to all of Europe, even though the Prussians and Austrians started sound they did not follow up their rewards, mainly when the road to Châlons and Paris lay open to both of them. At Valmy, the Prussians were beaten, and afterward were removed across the frontier by death and illness. Dumouriez at that time occupied Flanders and was triumphant over the Austrians at Jemappes. That was followed by the fall of

Ghent, Brussels, Antwerp, Mons, Malines, and Namur, even though less important towns, such as Tournay, Ostend, and Bruges welcomed the triumphant troops with open arms as the heralds of a new era.

In Savoy, the weak Sardinian forces were engaged by Montesquieu, what resulted in the country seized as well as Nice by Anselme. Overall the result of the new drawing of 1793 had Belgium also seized and the Dutch territories was also decided. This situation was a tremendously stupid move.

Holland and England

Holland and England became involved, when the French proclaimed their aggressive objectives to both powers. Prior to this point, Great Britain stayed strictly neutral and began proceedings to send 10,000 troops to support Holland. These act should make the difference, however England send the incompentent Duke of York to lead the troops. In addition to sending troups, the Hessioans and Hanoverians were paid by English gold.

The Island Kingdom and Russia had already allied themselves, although the Czarina's designs on Poland precluded close collaboration, and throughout the next months Naples, Prussia, Sardinia, Spain, the Holy Roman Empire, and Portugal joined in joint support against France.

Duke of Wellington

Dumouriez properly appeared on Dutch terrain but was forced to retreat on Flanders by the loss of the general engaged in blockading Maastricht. On resuming attacking operations, he lost the battle of Neerwinden. Within the next night, the French had abandoned all their invasions plans in Belgium, which once more handed over into the possession of Austria. Dumouriez took sanctuary in the camp of the Imperialists after negotiating with Coburg, the commander of the "White Coats," to place the border strongholds into his hands and to combine the two armies. None of these arrangements were carried out, because the crushed General found it more wise to leave the republic.

Mayence, a town on the Rhine, was advanced by the Prussians, to whom it in the end surrendered, and Valenciennes and Condé were successfully surrounded by the British and Austrians. In the end during July 1793, all three fortresses fell.

The tide was starting to turn for France. There were open revolts in Toulon and Lyon and shortly after civil war broke out in La Vendee. Had the Allies at that time made a concentrated effort, combined their forces, the downfall of the Republic might have followed. However they were all so full of themselves, instead of taking up their advantage. They were so concerned protecting, that the strenghth of the army was divided into groups. Much precious time was wasted on Dunkirk made by the English, Hessian, Hanoverian and some Australian forces, in total about 37,000 strong. The Revolutionary Government, expanding its fighting organization, ordered General Houchard to attack the rival forces at the historic seaport. The consequences of this move were, that the Duke of York was required to retreat and had to leave forty guns and much of his luggage behind. Houchard's victory was short-lived. He met with tragedy and paid the price of failure with his head. With the Pact defeat spelt death.

By the middle of September, all the vital strongholds which defended the passage way of the Allies troups to the Capital had fallen, with the exception of Maubeuge. The triumph of Jourdan, the beneficiary of Houchard, over the covering force at Wattignies saved the status quo, and on the 17th October the French marched into Maubeuge. On the Upper Rhine, the Allies found themselves in control of Metz only, at the end of 1793. In the south-west of Europe, the battle made no further progress, and the Republican cause increased fresh stimulus by the crushing of the royalist risings at Lyons and Toulon. It will be remembered that Napoleon won his first laurels in helping to conquer the enormous teritorry in the south of France, and in forcing the

Duke of Wellington

removal of the British fleet under Hood which had gone to backing the rebellion.

These realities, are may not be remembered as important, however they are crucial to come to a accurate understanding of Wellington in the early days of the Great War, that is detailed out in the next chapter.

CHAPTER II
Wellington's Debut (1794–97)

"I learnt what one ought not to do,

and that is always something."

Wellington.

Outline of specific point of views or inside knowledge in this chapter:

4. Similar events happened during the Peninsular War.
5. At Arnheim, on the Rhine, fewer than twenty-five miles distant. Giving to the de Ros MS., referred by Sir Herbert Maxwell, Dundas visit Wellesley "about once a fortnight."
6. Ireland's Lord-Lieutenant, 1794–8.
7. Letter to Sir Chichester Fortescue, dated 20th June 1796, quoted by Sir Herbert Maxwell, vol. i. p. 19 n.

The pages of military romance are filled with references to the disappointed lover who tries to calm his sadness by signing into active service. In real life, it is uncertain that such things often occurred, but it appears that it was accurate for Arthur Wellesley. He asked his oldest brother to use his power with Pitt to encourage Lord Westmorland to send him "as major to one of the border troups," his own regiment being "the last for service." The request was refused, and the young officer had to postpone until May 1794. Orders were then dispensed for the 33rd to advance on foreign service as part of a contingent under Lord Moira which was immediately required to strengthen the Duke of York.

The Allies had not only experienced a sequence of defeats, but Prussia felt back with many of her forces on the Rhine for service in Poland, the dividing into many divisions of which seemed to offer more concrete gains than the extended warfare against "armed opinions." As an associate of the Holy Roman Empire, Prussia had the requirement to supply 20,000 troops—a sheer handful—and she proclaimed her intention of just satisfying this obligation. Once more British gold came to the salvage, and Prussia, by a pact signed on the 19th April 1794, approved to keep 62,000 men at the disposal of the Allies in return for a substantial funding. The unfortunate Austrian general, Mack, was then given command of the new campaign. The lethal fault was repeated of dividing the army, with the consequence that whereas the Imperialists under Clerfait were required to retreat on Tournay, the Duke of York, assisted by Prince Schwartzenberg, secured a lead at Troisville. A series of movements around Tourcoing followed on the 16th to the 18th May, through which his Highness barely escaped being a prisoner, owed partly to his having been left secluded by the cutting off of his communications, and partly to a admirable strength of mind to hold the positions his troops had gained. At Pont-à-chin, near Tournay, the constant efforts of Pichegru to protected the village finished in disaster. On the 26th June the Austrians, in their attempt to release Charleroi, which had lay down his arms to the increasing forces of the French under Jourdan a few hours before, were required to withdraw from the plains of Fleurus. "The defeat of Flanders," says Alison, "directly followed a fight which an innovative general would have transformed into the most significant triumph." The Duke of York, having to fall back at Oudenarde, was also withdrawing, committed upon covering Antwerp and Holland.

Wellesley reached Ostend

Wellesley enterred into Ostend with his regiment in June 1794, from where he was directed to Antwerp, to which the Duke of York and the Prince of Orange soon fell back although Moira marched to Malines. The Colonel believed that his senior officer would have been better advised had he and

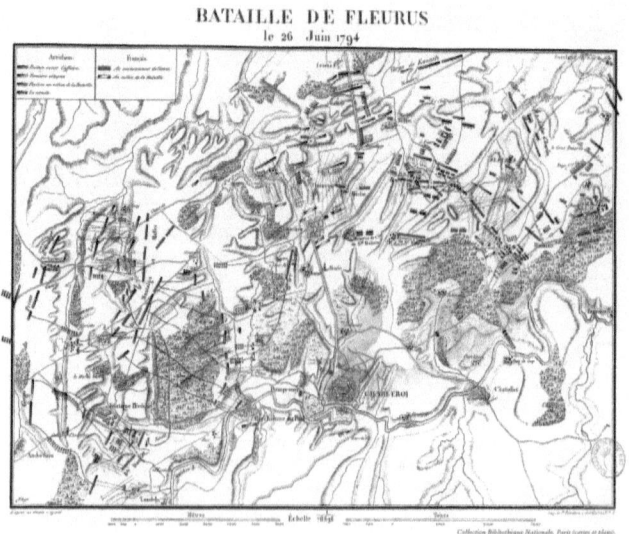

Figure 4 Map of Battle of Fleurus

his troops progressed up the Scheldt or the Maes in boats, a judgement afterward confirmed by events.

After settle down pressing troubles, Wellesley accepted his commands and reached the Duke of York several days before Moira was in touch with him. It was a moral victory for the young officer and doubtless served the very useful purpose of inspiring his desire.

For three months, the Duke of York and the Prince of Orange remained in Antwerp. The Commander of the Dutch troops then retired towards the Rhine, and the previous moved towards Holland. During the march General Abercromby was said to lock down the village of Boxtel, taken on the earlier evening by one of Pichegru's divisions. A desperate fight succeeded, and aside the heroic bravery of the British infantry, cavalry, and artillery, it concluded in disaster. It is very likely that the entire force would have been crushed but for Wellesley's promptitude in covering the withdrawal. No hostility was presented until the British were passing through a wood when a camouflaged battery opened fire. A little later there was substantial confusion, and a body of French Hussars advanced forward only to meet Wellesley's battalion pinched across the road. They were prepared, thanks to the bravery of the young commander.

Duke of Wellington

Throughout an extremely severe winter, the British were constantly pushed by the passionate Republicans. From October to January 1795 Wellesley detained a post on the Waal, and the difficult nature of his duties is described by him in letters written at the time. "At present," he says on the 20th December 1794, "the French keep us in a continuous state of alarm. We turn out once, occasionally twice, every night. The officers and men are stressed to death, and if we are not released, I am certain of there will be very few of the them remaining shortly. I have not changed my clothes for a long time, and largely spend the greatest part of the night upon the bank of the river, which helped me to completely got rid of that sickness which was near killing me at the end of the summer campaign. Although the French upset us much at night, they are very amusing during the daytime; they are continually chattering with our officers and soldiers, 4

and dance the carmagnole upon the opposite bank whenever we wish them, but sometimes the spectators on our side are broken up in the middle of a dance by a cannon ball from theirs."

It is a genial, kind message, but Wellesley always held his feelings well under control. In the above he chose to reveal the humorous aspect of the lengthy suffering. There was plenty to complain about what he had desired. The food supply was lacking; the wounded had to bear their sufferings because the stock of medicines ran short, and in general hardship was terrible. A sad lack of foresight characterized the whole campaign. What could be expected of a Commander-in-Chief who gave preference to the pleasures of the table if a dispatch arrived during a meal, and contemptuously remarked, "That will keep till the morning"? During the time of his sojourn on the Waal, Wellesley "only saw once one general from the headquarters, [5] which was old Sir David Dundas.... We had letters from England, and I declare that those letters told us more of what was passing at headquarters than we learnt from the command center ourselves.... It has always been a marvel to me how any of us escaped."

That "old Sir David Dundas" thought very highly of the young officer's conduct is evident. When he succeeded as Commander-in-Chief of the British forces, on the recall of the Duke of York in the following December, Wellesley was appointed Brigadier and given command of the rear guard. By a series of retreats, the tattered army eventually reached Bremen. It embarked for England early in 1795.

In summing up Wellesley's first experience of field service, Earl Roberts states that it was, "no doubt, treasured to Wellington in the years to come. It must have taught him that soldiers even of the best quality, well drilled, disciplined and equipped, cannot hope to be successful unless proper arrangements are made for their supply and transport. As well as unless those who direct the operations have formed some definite plan of action and have sufficient zeal and professional knowledge to carry it out. If the French generals had taken full advantage of the opportunities which the incapacity of the English and German commanders threw in their way, the British force must have been annihilated."

One is inclined to doubt whether the troops were "well drilled, disciplined and equipped" at this period. The gross incompetence of many of the highest officers is abundantly proved, and continued lack of success speedily reduces the fundamental strength of any regiment.

As already noted, the commissariat was execrable. We have it on the authority of one who was present that during the retreat hundreds of invalids succumbed, "while the shameful neglect that then pervaded the medical department, rendered the hospitals nothing better than slaughter-houses for the wounded and the sick."

Duke of Wellington

Wellesley reached England

And shortly after, he decided to leave the Army. The cause is unknown, but it seems greatly probable that either his recent experience had disgusted him with the service as constituted, or he wished to obtain more remunerative employment so that he might be in a position to marry the lady of his choice. He also owed money to his brother, who had made advances for his promotion. This sum could be repaid by the sale of his commission. Although Wellesley was always scrupulous in money matters, the reason seems scarcely credible. We are therefore forced to accept one of the other alternatives, perhaps both, for mention is made of the miserable state of the Army in his letter to Lord Camden [6] regarding the desired appointment. He consulted Mornington on the matter, and it was decided that a position under the Revenue or Treasury Boards would serve his purpose. "If your Excellency," he writes to the Viceroy, "is of the opinion that the offices at these boards are too high for me, of course, you will say so. I am convinced that no man is so bad a judge of a claim as he who makes it, I trust you will not believe that I shall feel otherwise towards you then as I have always felt, with sentiments of the greatest regard.... You will probably be surprised at my desiring a civil instead of a military office. It

Duke of Wellington

is certainly a departure from the line which I prefer, but I see the manner in which the military bureaus are filled, and I don't want to ask you for that which I know you cannot give me."

Research has failed to discover what answer, if any, was vouchsafed this communication. Wellesley remained in the Army. In October 1795, he and his regiment sailed from Southampton as part of an expedition against the French settlements in the West Indies. The vessels encountered a terrible storm, still known as "Christian's Storm," after the name of the admiral who commanded the fleet. While it might be untrue to say that the ships were in an unseaworthy condition, their sanitary state was deplorable, for they had but recently returned from a lengthy voyage as hospital and prison transports. Scarcely forty-eight hours after they had sailed, and when they were off Weymouth, the full force of the blast struck them. One vessel foundered with all hands, half-a-dozen or more were totally dismasted, and hundreds of soldiers went to their death in a battle with the elements against which all the drill in the world was ineffectual. Fortunately, Wellesley escaped, but when he received orders, in April 1796, to embark his men for India, he was too ill to accompany them. However, he set sail for Calcutta in June, and overtaking the 33rd Regiment at the Cape of Good Hope, duly reached his destination in February 1797. "The station is so highly advantageous to him that I could not advise him to decline it," says Lord Mornington.[7] The good-natured Earl little knew what advantage, both to Wellesley and the Empire, was to accrue as the result of the failure of his brother's political ambitions.

Duke of Wellington

CHAPTER III
Battle of Seringapatam
(1797–1800)

India, "a country fertile in heroes and statesmen."

Canning.

Outline of specific point of views or inside knowledge in this chapter:

8. Sir Herbert Maxwell, p. 35.
9. Gleig (p. 26) says £7000, Roberts (p. 11) £7000 in money and £1200 in jewels. Sir Herbert Maxwell (p. 39) calls attention to a letter, dated the 14th June 1799, in which Wellesley "gives it as 3000 pagodas in jewels, and 7000 in money; in all, 10,000 pagodas, equal to about £4000."
10. In later years, Wellington offered to provide for the unfortunate Spanish general, Alava and gave him a small house in the park of Strathfieldsaye.

The proverb to the effect that "History repeats itself" is not strictly true. The further we study the subject, the more we find that like causes do not necessarily bring about similar effects. The ill success which attended the expedition to the West Indies, ere it left the English Channel, has a fitting parallel so far as its practical utility is concerned in the force placed at General St Leger's disposal to attack Manilla. The Philippine Islands then being in the possession of Spain, with whom Great Britain was now at war. Fortunately, it did not meet with disaster, but neither expedition reached its destination. Wellesley accepted the offer of Sir John Shore, the Governor-General of India, to command a brigade, and the troops were embarked. They had not proceeded farther than

Duke of Wellington

Penang before an order was issued for their recall owing to troubles brewing in India itself.

Return to Calcutta

Shortly after his return to Calcutta the Colonel was placed in command of the forces in Madras. He also heard that his eldest brother had been offered the extremely responsible and agreeable post of Governor-General in succession to Sir

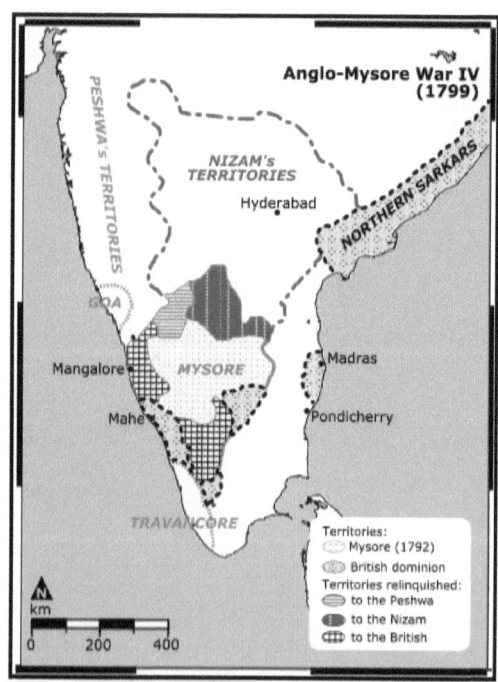

Figure 5 Anglo-Mysore_War

John Shore. It was now his turn to feed the flames of Mornington's ambition. He writes: "I strongly advise you to come out. I am convinced that you will retain your health; nay, it is possible that its general state may be improved, and you will have the fairest opportunity of rendering material service to the public and of doing yourself credit." Mornington lacked self-confidence, and a thousand and one doubts and fears possessed his mind. The Colonel reminded him that if he refused so advantageous a position on account of his young family, "you forego both for yourself and them what will certainly be a material and lasting advantage."

Mornington accepted, and arrived in Calcutta with his youngest brother, Henry, as private secretary in the middle of May 1798. He speedily found an antidote for homesickness in endeavoring to unravel the tangled skein of affairs in Mysore, where Tipú Sultan was intriguing with the French Republic for assistance in attacking the possessions of the East India Company in Southern India. The pugnacious character of the son of Hyder Ali was typified by the tiger's stripes on his flag. He possessed the fanaticism and barbarity of the Oriental at his worst, and when the opportunity occurred would feed a beast of prey with an English prisoner.

Duke of Wellington

WELLINGTON'S BATTLE IN INDIA.

To secure either the friendship or the neutrality of the Nizám, whose territory abutted that of the bloodthirsty Tipú now became of paramount importance. His army was officered by Frenchmen, which was proof positive that in the event of war it would assist Britain's enemy although the Nizám had a distinct leaning towards the English. As it happened, the native troops mutinied against their officers, and, seizing his opportunity, the Nizám dismissed them. They were sent to England as prisoners and subsequently allowed to return to their own country, a most humane consideration, for which Mornington was mostly responsible. The military positions they formerly occupied were promptly filled by our own officers. A new treaty was made to preclude the Marhattás from allying themselves with Tipú, and a force of 6000 British troops was maintained by the Nizám at Hyderabad.

Meanwhile Wellesley had proceeded with his regiment to Madras, and, owing to the death of the senior officer, was placed in temporary command of the troops. In communication with Lord Clive, the Governor of the Presidency, and General Harris, the Commander-in-Chief, he busied himself with the multitudinous arrangements necessary for an advance upon Seringapatam, the capital of the Mysore

Dominions. Horses, bullocks, and elephants had to be provided for the purpose of transport; forts equipped and provisioned; the siege train properly organized. He drew up a plan of campaign and bent himself to the task with exacting energy. Notwithstanding the preparations for war, he still hoped that a resort to arms would prove unnecessary. Those who are apt to think that all military men delight in strife for the sheer love of it will do well to remember this fact and judge less harshly, for Wellington is the typical representative of the British Army. But he believed in being ready, and hated nothing so much as "muddling through."

There was still a possibility, though scarcely a probability, that Tipú would repent. He had received word that Napoleon, then on his famous Egyptian expedition, was coming to his aid with an "invincible army." So far he had refused a definite statement of policy. Not until it was abundantly evident that the protracted negotiations with the Sultan of Mysore with the Government were mere to gain time, was a declaration of war issued on the 22nd February 1799. According to Wellesley, General Harris "expressed his approbation of what I had done, and adopted as his own all the orders and regulations I had made, and then said that he should mention his consent publicly, only that he was afraid others would be displeased and jealous. Now as there is nothing to be got in the army except credit, and as it is not always that the best intentions and endeavors to serve the public succeed, it is hard that when they do succeed they should not receive the approbation which it is acknowledged by all they deserve. I was much hurt about it at the time, but I don't care now, and shall certainly do everything to serve General Harris, and to support his name and authority."

Wellesley never feared to speak his mind, as his many dispatches abundantly testify. In a letter to Mornington, he admits that he had "lectured" the Commander-in-Chief because he allowed the Madras Military Board too much license in the matter of appointments. On the other hand, he had "urged publicly to the army (in which I flatter myself I have some influence) the necessity of supporting him, whether he be right or wrong." In his opinion, it was "impossible" to hold the General "too high if he is to be the head of the army in the field."

Harris surely compensated Wellesley to some extent by placing him in command of thirteen regiments, including the Nizám's contingent, with the rank of brigadier. The strength of this force was about 16,000 men, that of the whole army 35,000, excluding 120,000 camp followers, the bugbear of the old-time commander. The Bombay corps under General Stuart attacked a portion of the enemy, commanded by the wily

Tipú, in the vicinity of Sedasser, on the 6th March. This success augured well, for the Sultan was forced to retire.

Harris's first serious engagement

Took place near Malavelly on the 27th, Wellesley advancing to the attack and turning Tipú's right flank. After an engagement lasting three hours the enemy withdrew, with the loss of some 2000 men by death or wounds against the British 7 killed and 53 wounded. Tipú was a skillful soldier and had not neglected to throw up a line of entrenchments before Seringapatam, into which city he now withdrew. To drive in the first outposts before definitely besieging the place was Harris's first object. This duty was entrusted to Wellesley and Colonel Shaw respectively, each having charge of a detachment. He failed, for reasons explained in the following letter:

"On the night of the 5th, we made an attack on the enemy's outposts, which, at least on my side, was not fairly so prosperous as could have been wished. The fact is, that the night was very dark, that the enemy expected us and were strongly posted in an almost impenetrable jungle. We lost an officer, killed, and nine men of the 33rd wounded, and at last, as I could not find out the post which it was desirable I should occupy, I was obliged to desist from the attack, the enemy also having retired from the post. In the morning, they re-occupied it, and we attacked it again at daylight and carried it with ease and with little loss. I got a slight touch on the knee, from which I have felt no inconvenience, and I have come to the determination never to suffer an attack to be made by night upon an enemy who was prepared and strongly posted, and whose posts had not been reconnoitered by daylight." It should be added that twelve soldiers were taken the prisoner and executed by the brutal method of nails being driven through their heads and that Wellesley had previously given it as his opinion that the projected attack on the grove would be a mistake. The operation undertaken by Colonel Shaw was successful.

The siege

Now proceeded in earnest, but a breach was not made in the solid walls surrounding Seringapatam for three days. On the 4th May, the place was stormed by General Baird. General Sherbrooke's right column was the first to ford the

Cauvery River. His men speedily scaled the ramparts and engaged that part of the Sultan's 22,000 troops stationed in the immediate vicinity.

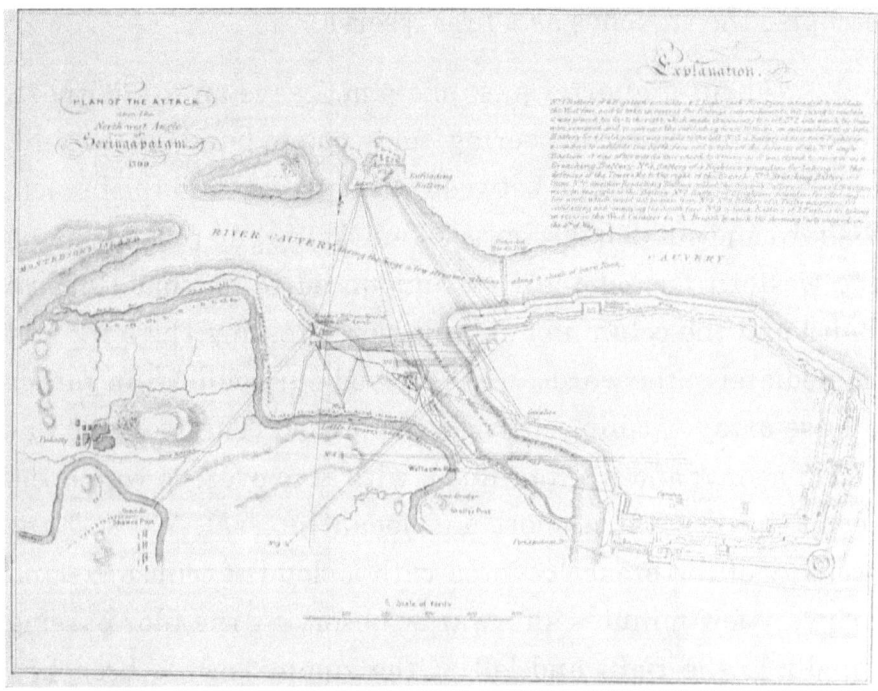

Figure 6 The plan of the Siege Saringapatam 1799

The defenders fought with the fatalistic energy and determination so characteristic of the natives of India. The left column followed but found the way more difficult. Tipú, mounting the ramparts, fired at the oncoming redcoats with

Duke of Wellington

muskets handed to him by his attendants. It was his last battle; his body was afterwards discovered in a covered gateway, together with hundreds of others. Wellesley, with his corps, occupied the trenches as a first reserve.

"About a quarter past one p.m.," says an eye-witness, "as we were anxiously peering, telescope in hand, at the ford, and the neutral ground between our batteries and the breach. A sharp and sudden discharge of musketry and rockets, along the western face of the fort, announced to us that General Baird and the column of assault were crossing the ford; and immediately afterwards, we perceived our soldiers, in rather a loose array, rushing towards the breach. The moment was one of agony; and we continued, with sore eyes, to watch the result, until, after a short and mean interval. We saw the acclivity of the breach covered with a cloud of crimson,—and in a very few minutes afterwards, observing the files passing rapidly to the right and left at the summit of the breach, I could not help exclaiming, 'Thank God! the business is done.'

Figure 7 Tipu's death

"The firing continued in different parts of the place until about two o'clock, or a little afterwards. When, the whole of the works being in the possession of our troops, and the St George's ensign floating proudly from the flagstaff of the Southern cavalier, announced to us that the triumph was completed."

On the 5th, Wellesley took over the command from Baird, who had requested a temporary leave of absence, and without delay began to restore some kind of order among the

British troops, whose one object after victory was plunder, in which matter they showed little delicacy of feeling. The city was on fire in several places, but the flames were all extinguished within twenty-four hours, and the inhabitants were "retiring to their homes fast." Having stopped, "by hanging, flogging, etc.," the insubordination of the troops and the rifling of the dead by the camp followers who had flocked in, Wellesley proceeded to bury those who had fallen.

Figure 8 Finding_the_body_of_Tippoo_Sultan_-_Samuel_William_Reynolds

Duke of Wellington

During the four weeks of the siege, the British lost 22 officers and 310 men, and no fewer than 45 officers and 1164 men were reported as wounded and missing. [8]

The Commander mentions that jewels of the greatest value, and bars of gold, were obtained. As the prize agents assessed the treasure taken at £1,143,216, the wealth of Seringapatam must have been astounding. Wellesley's share came to about £4000. [9]

Hundreds of animals were required to carry the opulent stuff, plate, and richly-bound books from this city of opulence. A little comic relief to so much sordidness is afforded by Wellesley's difficulties regarding some of the late Sultan's pets. "There are some tigers here," he writes, "which I wish Meer Allum would send for, or else I must give orders to have them shot, as there is no food for them, and nobody to attend to them, and they are getting violent." Tipú's 650 wives gave less trouble than the wild beasts. They were removed to a remote region and set at liberty.

Duke of Wellington

Commander of Mysore

Wellesley's next appointment was as Commander of the Forces in Mysore. He proved himself to be particularly well fitted for the post, which apparently required a man of infinite tact, who could be lenient or severe as circumstances demanded. It was Wellesley's testing time, and he did not fail either in administration or the rough and tumble of the "little war" so soon to fall to his lot. He had already served on a commission appointed to go into the question of the partition of the conquered Dominions, a small part of which was made over to the Peshwá, and larger shares of the Nizám and the East India Company respectively. The dynasty overturned by Tipú's father was restored. As the new Rájá of Mysore was

Figure 9 Commander of Mysore General Wellesley's House

only five years of age, he was scarcely able to appreciate the fact that his territory was so greatly diminished.

King of the World

We now come to a story worthy of a place in the Arabian Nights. It concerns an adventurer who, later, assumed the truly royal title of King of the World. Dhoondia Waugh, to give him the name by which those who were unfortunate enough to make his acquaintance first knew him, was the chief of a band of robbers whom Tipú had captured and thrown into prison. Recognizing in him a brave man, the Sultan remitted the sentence of death and gave him a military appointment, thus turning his acknowledged abilities into a less questionable channel, for a thief must need to be fearless and daring if he is to succeed. For some reason not altogether clear, Dhoondia Waugh was again imprisoned, and he did not regain his liberty until the fall of Seringapatam, when he was liberated, together with some other gaolbirds. The old thieving instinct reasserted itself, and as he encountered no difficulty in collecting a band of the late Tipú's cavalry, he speedily resorted to means and measures which alarmed the inhabitants of every place he visited. When pressed by the troops sent after them the horde took refuge in the territory of the Peshwá, the nominal head of the Marhattá confederacy.

Duke of Wellington

There they received anything but a cordial welcome, although it seems probable that reinforcements were obtained among the malcontents. However, that may be, Dhoondia Waugh duly appeared near Savanore. Having the safety of the Mysore Dominions very much at heart, for he had supreme civil and military control, Wellesley started in pursuit of the freebooter. Several fortresses held by Dhoondia's unlawful bands were stormed, his baggage taken, and some guns captured.

An affray which took place near the Malpurda River at the end of July 1800. It not only reduced the chief's forces but caused many of his followers to forsake the cause, although their strength in the following September was considerably more than that at Wellesley's command; in actual figures, some 5000 against 1200. The operation on the 10th of that month, which proved decisive, was rough, for the enemy was strongly posted at a village called Conahgull. The Colonel charged with such cool daring and so determined a front, that after having stood firm for some time the enemy made off, closely pursued many miles by the British cavalry. A dire and just retribution was exacted; those who were not killed "were scattered in small parties over the face of the country." The King of the World had fought his last battle. He was found among the slain.

Duke of Wellington

It is frequently asserted that Wellesley held but a stumpy opinion of the troops which he commanded, and he certainly passed harsh judgment on those who shared his later campaigns. Not so in this particular instance, however. In the dispatch detailing "the complete defeat and dispersion" of the forces of Dhoondia, he expressly remarks on the "determined valor and discipline" of the soldiers, the patience and perseverance displayed in "a series of fatiguing services," and the excellent organization of the commissariat department.

Wellesley also showed that a kind heart is not necessarily the attribute of a weak nature. With a humanity entirely worthy so great a man, he had Dhoondia's "supposed or adopted son" cared for, and afterwards placed £400 in the hands of trustees for his future use. **10**

"Had you and your regicide army been out of the way," writes Sir Thomas Munro to Wellesley, "Dhoondia would undoubtedly have become an independent and powerful prince, and the founder of a new dynasty of cruel and treacherous Sultans."

This short campaign likewise furnishes us with one of the secrets of the success of our national military hero. Just before he set out on the long chase after the King of the World, he was offered a position particularly rich in prospects, namely, the military command of an expedition for the surrender of the Dutch island of Batavia. The sole condition was that Lord Clive, the Governor of Madras, to whom he was responsible, could spare him. A man who was moved by purely personal ambition would have had no hesitation in bringing all his influence to bear on the Governor to secure so worthy an opening. Wellesley, however, recognizing that he had already begun preparations for the running to earth of the bloodthirsty and cruel Dhoondia—an end much to be desired—asked Clive to accept or decline for him as he thought best. He neither pleaded for nor against, although he hoped that if Admiral Rainier were not starting at once he might be able to join him when the work on hand was finished. "I am determined that nothing shall induce me to desire to quit this country until its tranquility is ensured. The general want of troops, however, at the present moment, and the season, may induce the Admiral to be desirous to postpone the expedition till late in the year. In that case, it may be convenient that I should accompany him...."

The Governor of Madras refused his permission, and there the matter ended. Months afterwards, when there seemed a probability of operations in the Marhattá Territory, Wellesley wrote a lengthy Memorandum on the means of carrying such a campaign to a successful issue. "The experience," he notes in his opening remarks, "which has been acquired in the previous contest with Dhoondia Waugh, of the seasons, the nature of the country, its roads, its produce, and its means of defence will be of use in pointing them out."

Thus, it will be seen that the knowledge gained by Wellesley during the performance of individual duty was stored up for future use. A march or a campaign was not basically carried out and then dismissed. It was a lesson learned and to be remembered. In military matters, he was to a very appreciable extent self-taught.

No drill-book in existence can furnish skill or assure victory, and genius itself is valueless on the battlefield without a clear perception based on things ascertained "the experience which has been acquired" referred to in the above communication. Napoleon, against whom Wellesley was to fight in the years to come, early recognized the supreme importance of this principle. "The gifted man," he says, "profits by everything, neglects nothing which can increase his chances."

CHAPTER IV
War with the Marhattás (1801–3)

"We must get the upper hand, and if once we have that, we shall keep it with ease, and shall certainly succeed."

Wellington.

Outline of specific point of views or inside knowledge in this chapter:

11. Created 20th December 1800.
12. The Austrian general, Kray, had succeeded Archduke Charles as Commander-in-Chief of the Army in Germany in the campaign of 1800, but owing to his ill-success he was superseded in a few months by Archduke John, hence Wellesley's reference.
13. "The Life of Wellington," pp. 45–6.
14. "Dispatches," vol. ii. p. 312.
15. "The Life of Arthur Duke of Wellington," by G. R. Gleig, M.A., F.R.G.S. (London Ed. 1864), pp. 33–4.
16. 79 officers and 1778 soldiers were killed and wounded. —Sir Herbert Maxwell, p. 58.
17. Gleig, pp. 37–8.

Duke of Wellington

That disappointments are frequently blessings in disguise had already been proved by Arthur Wellesley. Unfortunately, it is easier to forget such a precept than to practice it, and each apparent failure to climb another rung of ambition's ladder is apt to be regarded as a definite setback. It was so with Wellesley, and a time of trial and perplexity followed the campaign of Seringapatam and the defeat of Dhoondia.

He eventually weathered the storm of depression which pressed upon him, as he weathered many another, but it must be admitted that he bent before it. It came about in this way. The French army in Egypt was still very active, although Napoleon had long since left it. He was now First Consul, and gradually preparing himself and the nation for the assumption of the Imperial crown. The Governor-General, henceforth to be known as Marquis Wellesley, 11 was of the opinion that a small expedition should be sent either to Batavia or the Mauritius or to assist Sir Ralph Abercromby in his attempt to drive the French out of Egypt.

With one of these desirable objects in view, his brother Arthur was given 5000 troops. He at once set off for Trincomalee, on the island of Ceylon, the headquarters of the little army, intent on personally superintending the arrangements. Shortly afterwards instructions came to hand from the Home

Duke of Wellington

Government that 3000 men were to be sent to Egypt. Colonel Wellesley was informed of this decision and determined to lose no time in forwarding the project. Without receiving official word to do so, and still believing he held the premier post, he embarked the men and sailed for Bombay, where he had ordered an ample supply of provisions to be ready.

When off Cape Comorin, Wellesley received a letter from his brother, stating that he had appointed Major-General Baird to the command of the troops destined for the island of Batavia, which made it clear that the Governor-General had not then received the dispatches of the Secretary of State. Knowing that some at least of the troops on the transports would be required for Egypt, he proceeded on his way and wrote to Baird of his intention. A little later the other letter came to hand from another source; but the fleet was in want of water, some of the troops had died, and "I was induced to adhere to my original plan."

Baird, who, on arriving at Trincomalee, found "the cupboard was bare," was deeply incensed at Wellesley's high-handed behavior. The "culprit's" feelings as to the Governor-General's new appointment were also far from Pacific. That he acted in perfect good faith is evident from the preceding, which is borne out in a lengthy dispatch in which he sought to justify his action in the eyes of his brother.

Duke of Wellington

"I have not been guilty of robbery or murder," he writes to Henry Wellesley from Bombay on the 23rd March 1801. "He has indeed changed his mind. The world, which is always good-natured towards those whose affairs do not exactly prosper, will not, or rather does not, fail to suspect that both, or worse, have been the occasion of my being banished, like General Kray, to my estate in Hungary. [12] I did not look and did not wish, for the appointment which was given to me. I say that it would probably have been more proper to give it to somebody else; but when it was given to me, and a circular written to the governments upon the subject, it would have been fair to allow me to hold it till I did something to deserve to lose it.

"I put special considerations out of the question, as they ought and have had no weight in causing either my original appointment or my supersession. I am not somewhat satisfied with the manner in which I have been treated by Government upon the occasion. However, I have lost neither my health, spirits, nor temper in consequence thereof. But it is useless to write any more upon a subject of which I wish to retain no remembrance whatever."

Baird would have been scarcely human had he not felt hurt by finding himself head of a force which had disappeared, especially as the Colonel had already superseded him as Governor of Seringapatam. But he forgave if he did not forget, and neither did Wellesley. Some thirty years afterwards, when Baird's days of active soldiering were over, he remarked, during a chat with Sir John Malcolm, who had himself done good service in India: "Times are changed. No one knows so well as you how severely I felt the preference given on several occasions to your friend Wellesley, but now I see all these things from a far different point of view. It is the highest pride of my life that anybody should ever have dreamed of my being put in the balance with him. His name is now to my joy, and I may almost say glory."

It is satisfactory to know that Arthur Wellesley was not foolish enough to allow the iron to enter into his soul to such an extent as to prevent him from co-operating with Baird. Into whose hands he placed a "Memorandum on the Operations in the Red Sea," accompanied by a letter acknowledging "the kind, candid, and elegant manner in which you have behaved towards me." When the expedition was ready, Arthur Wellesley was laid low with a fever. Consequently, the Commander-in-Chief was obliged to sail without his lieutenant, not altogether to his discomfiture one would surmise.

An attack of the dreaded Malabar itch did not tend to a speedy recovery of the invalid, but he was sufficiently well in May 1801 to resume his former duties at Seringapatam, where he had been reinstated by his brother. By living moderately, drinking little or no wine, avoiding much medicine, taking exercise, and keeping his mind employed, he eventually recovered. As Baird saw no fighting, his rival lost nothing by remaining in India.

Sir Herbert Maxwell assumes that Arthur Wellesley's fever was caused by disappointment. In reality as the latter expressly states that Baird's "conduct towards me has by no means occasioned this determination (namely, to resign the appointment), but that it has been perfectly satisfactory," the statement is evidently based on a surmise that the Colonel was diplomatically lying. Everybody fully appreciates the influence of mind over matter, and thwarted desire may have weakened Wellesley's health, but surely the facts of the case scarcely justify so bold an assertion.

Colonel Wellesley remained in Mysore

For nearly two years, during which he did his work both wisely and well, showing favor to none and justice to all. It was in February 1803 that the forthcoming Wellington, now a Major-General received news that he was required for51 active service against the Marhattás. The war-like intentions of this powerful confederacy, which alone could challenge British supremacy, had not escaped the notice of Government. The nominal head of the five native princes who constituted it was Baji Rao, the Peshwá of Poona, the others being Daulat Rao, Sindhia of Gwalior; Jeswant Rao, Holkar of Indore; the Gaikwár of Baroda, and the Bhonsla Rájá of Berar. Sindhia was the most powerful and possessed a superior army drilled by French officers and commanded by Perron, a deserter from the French Marine.

Holkar had at his disposal no fewer than 80,000 splendidly-equipped men, mostly cavalry, likewise organized by European soldiers. Intense rivalry existed between these princes, and when, in October 1802, the latter invaded Poona, the armies of Sindhia and the Peshwá met with disaster. The Peshwá sought refuge with the British, and forthwith entered into an offensive and defensive alliance with Lord Wellesley as the only means of saving his territory. The key clauses were that 6000 British troops should be kept at Poona. The expense being met by the assignment to the East India Company of a particular territory; that the Peshwá would not make war with the other princes or allow them to prey on each other without the consent of Government; and that he should be reinstated in his capital. This arrangement, known as the Subsidiary Treaty of Bassein, soon had the effect of drawing together the remaining members of the Marhattá confederacy, cementing a friendship between Sindhia and Holkar, and an alliance between Sindhia and the Bhonsla Rájá. It is clear that the continued acknowledgment of the Peshwá as head of the confederacy, now that he was under the ægis of the British, would have been to admit the supremacy of the conquering Power they so much resented. Lord Wellesley had already signed a defensive alliance with the Gaikwár of Baroda. To be ready for eventualities, men from the armies of the three Presidencies, namely, Madras, Bombay, and Calcutta, were concentrated at various points. The first for operation on the north-west frontier of Mysore, the second for action about Surat and Broach, and the third for the occupation of Cuttack. A large force was also ordered to assemble at Cawnpore under General Lake, the Commander-in-Chief in India while three corps were held in reserve. Major-General Wellesley was placed in command of a detachment of some 10,600 troops, to which must be added the Nizám's contingent of 8400 men under Colonel Stevenson, making 19,000 in all. His orders were to secure Poona, now held by a small garrison of Holkar's soldiers totaling not more than 1500. He was already on the march when he heard of the intention of the Governor, acting on Holkar's instructions, to burn the town on the approach of the British.

"We were within forty miles of the place"—Wellesley himself tells the story [15] — "when this resolution of Holkar's lieutenant was communicated to me. My troops had marched twenty miles that day under a burning sun, and the infantry could no more have gone five miles farther than they would have flown. The cavalry, though not fresh, were less knocked up, so I got together 400 of the best mounted among them and set off. We started after dark on the night of the 19th of April, and in the afternoon of the 20th we got close to the place. There was an awful uproar, and I expected to see the flames burst out, but nothing of the kind occurred. Amrut Rao—that was the Marhattá's name was too frightened to think of anything except providing for his own safety, and I had the satisfaction of finding, when I rode into the town, that he had gone off with his garrison by one gate as we went in by another. We were too tired to follow, had it been worthwhile to do so, which it[53] was not. Poona was safe, and that was all I cared for." In the following month, the Peshwá returned to his capital.

Sindhia and the Rájá of Berar now busied themselves with gathering a large army at Burhanpur, ready to threaten the Deccan, Holkar retiring to Indore. Wellesley was no less active at Poona; his experience in Holland had taught him the fundamental lesson that an efficient organization is a powerful ally. Also, he was busy endeavoring to come to terms with Sindhia and the Rájá, for which purpose he had been given chief command of the British forces in the Marhattá states, with the fullest political authority. Similar powers were vested in General Lake in Northern India. After wasting as much time as possible in the negotiations so as to gain it for military preparations, Wellesley anticipated the inevitable. "I offered you peace on terms of equality," he writes on the 6th August 1803, "and honorable to all parties: you have chosen war, and are responsible for all consequences." On the following day, hostilities were declared against Sindhia and the Rájá of Berar.

The stone-built fortress of Ahmednuggur

The capture of which would safeguard his communications with Poona and Bombay and prevent reinforcements from Southern India reaching the enemy was his first object of attack. The main body of Sindhia's men was threatening Hyderabad, but the place was well garrisoned and so solidly constructed that it looked as though it would defy whatever artillery could be brought to bear on it. Wellesley said that except Vellore, in the Carnatic, it was the strongest country fort he had ever seen. However, he began operations against the outworks on the 8th, after having made proposals for its surrender without favorable result. "The Arabs," we are told, "defended their posts with the utmost obstinacy," but towards evening were forced to quit the wall. On the following day, the ground in the neighborhood of the fort was reconnoitered and a commanding position seized, on which a battery of four guns was constructed for use during the attack. The first shots were fired on the 10th at dawn, and the storming party speedily began its work. Three times an officer ascended a scaling ladder propped against one of the walls, and thrice he was hurled down by the defenders. The fourth attempt was successful, and, followed by some of his men, the gallant soldier literally hewed away into the town. The remaining troops, pressing on, took the place of those who fell. At length the Commander of the enemy's forces surrendered, "on condition that he should be allowed to depart with his garrison, and that he should have his private property." His fourteen hundred men marched out of the fort, and Wellesley's troops took possession.

On the 23rd September the General found himself and his small contingent of some 8000 soldiers face to face with the whole combined army of Sindhia and the Rájá of Berar, a state of affairs brought about by unreliable information, causing the separation of Wellesley and Stevenson. At least 50,000 of the enemy were posted in a strong position behind the river Kaitna, near the village of Assaye. As Wellesley had received no reinforcements and had only 17 guns compared with 128 commanded by skillful French officers at the disposal of the Marhattás, the disproportion of the forces was sufficiently understandable. To a general less experienced or daring the situation would have been considered sufficient cause for an instant retreat; even he called the attack "desperate." The problem for him to settle was, should he wait a few hours for Stevenson, or begin immediately with the limited resources at his disposal? Although only 1500 of his men were British, the Commander-in-Chief decided on the latter alternative, ignoring the information vouchsafed by his guides that the river was absolutely impassable. Yet it was only by crossing the stream that he could take advantage of the opportunity to attack. Here Wellesley's native wit and acute intelligence—he called it "common sense"—assisted him. His telescope merely revealed a village on either side of the stream. This fact suggested the probability of a neighboring ford. On investigation, such proved to be the case, and if the passage was challenging the General was at least fortunate in being able to carry out the operation without severe molestation by the enemy, who had foolishly neglected to guard this point. They repaired the omission so far as was possible by firing upon the advancing army as it slowly waded across, but the losses were comparatively trivial. "All the business of war," Wellesley once told Croker, "and indeed all the business of life is to endeavor to find out what you don't know by what you do."

The battle began well by the routing of some of the infantry and artillery by the Highlanders and Sepoys. This advantage was almost immediately counterbalanced by the mistaken zeal of the officer commanding the pickets, supported by the 74th Regiment. He foolishly led his men against the village, thereby exposing them to the concentrated fire of the enemy's artillery and musketry stationed there. Had he taken a less direct route, this could not have happened, but his enthusiasm overruled his caution. Men dropped down like ninepins in a skittle-alley when the ball is thrown by a skillful player. They fell by the dozen as they came within the zone of fire. Their comrades filled up the tell-tale gaps and continued to push on with a dogged tenacity entirely worthy their intrepid commander. Meanwhile what few British guns remained pounded away, and were silenced one by one as the men who worked them fell dead at their post. The enemy's cavalry then proceeded to decimate the already sorely depleted ranks of the 74th.

At this moment the 19th Light Dragoons, under Colonel Maxwell, were hurled at Sindhia's troops. The charge turned the fate of the day. What remained of the 74th rallied to the support thus given, and when Wellesley led the 78th into action, the village fell. An attempt was made by the enemy to rally, but it was too late. Men who, with true Oriental cunning, had fallen as though killed to avoid the oncoming British cavalry as they charged, and had escaped the iron-shod hoofs of the horses, rejoined the ranks, only to find that the day had been lost. The whole body was soon flying helter-skelter from the blood-stained field towards Burrampur, abandoning artillery, baggage, ammunition—everything that precluded swift movement. Twelve hundred of the Marhattás breathed they're last on this memorable day.

In fighting this battle "the hardest-fought affair that ever took place in India"—o'er again in the twilight of his days, the Duke of Wellington made light of the indiscretions of the officers at Assaye and remembered only their bravery. "I lost an enormous number of men: 170 officers were killed and wounded, and upwards of 2000 non-commissioned officers and privates; **16** but we carried all before us. We took their guns, which were in the first line, and were fired upon by the gunners afterwards, who threw themselves down, pretending to be dead, and then rose up again after our men had passed; but they paid dearly for the freak. The 19th cut them to pieces. Sindhia's infantry behaved admirably. They were in support of his cannon, and we drove them off at the point of the bayonet. We pursued them as long as daylight lasted and the exhausted state of the men and horses would allow; and slept on the field." 17

Figure 10 Fort of Beharampur West Bengal c1850

Wellesley himself, although not wounded, lost two horses. An eye-witness has recorded that he had never seen "a man so casual and collected as he was the whole time." Stevenson arrived on the following evening and set out almost immediately to follow the enemy, Wellesley being forced to remain owing to his lack of transport for the wounded, whom he refused to leave. The Colonel seconded Wellesley's magnificent victory by reducing the fortress of Burrampur on the 16th October, and that of Asseerghur on the 21st. Wellesley covered Stevenson's operations and defended the territories of the Nizám and the Peshwá. "I have been like a man who fights with one hand and defends himself with the other," he notes on the 26th October. "I have made some terrible marches, but I have been remarkably fortunate. First, in stopping the enemy when they intended to press to the southward, through the Casserbarry ghaut; and afterwards, by a rapid march to the northward, in stopping Sindhia, when he was moving to interrupt Colonel Stevenson's operations against Asseerghur; in which he would otherwise have undoubtedly succeeded."

Duke of Wellington

CHAPTER V
Last Years in India (1803–5)

"Time is everything in military operations."

Wellington.

Outline of specific point of views or inside knowledge in this chapter:

18. Representative.
19. Alison in his "Lives of Lord Castlereagh and Sir Charles Stewart" (vol. i. p. 175), says that it generally took six months to make the voyage. When Sir James Mackintosh sailed from Portsmouth for Bombay in 1804 his vessel only occupied three months and thirteen days (see his "Memoirs," vol. i. p. 207).
20. "His relationship to the Governor-General naturally lent much weight to his views with Lord Clive and General Harris, but," Sir Herbert Maxwell adds (p. 24), "it is remarkable how freely and frequently the elder brother sought, the younger's advice."
21. "The Life and Correspondence of the Right Honble. Henry Addington, first Viscount Sidmouth," by the Honble. George Pellew, D.D. (London, 1847), vol. ii. p. 242. In this connection see also "Wellington's Despatches," vol. ii. pp. 335–36 n., and "Despatches,

Minutes, and Correspondence of Marquess Wellesley, K.G.," vol. iii. p. 543.
22. "The Rise of Wellington," by Earl Roberts, V.C., p. 26.
23. "Personal interest was as much recognized in those days as the primary motor in the military promotion, as seniority and merit are now."—Sir Herbert Maxwell, vol. i. p. 67.

Bhonsla Rájá now became the immediate object of Wellesley's attention. While proceeding in quest of him the General received envoys from Sindhia requesting an armistice. This was granted on the 23rd November 1803, the prime condition imposed by Wellesley being that the enemy's army should retire forty miles east of Ellichpúr. This clause was not fulfilled, the cavalry of the wily Sindhia encamping at Sersooly, some four miles from the position occupied by Manoo Bappoo, brother of the Rájá, ready for immediate co-operation. Having again united their divisions, Wellesley and Stevenson pushed towards them. "A confused mass" about two miles beyond Sersooly proved to be the enemy's armies on the march. A little later the General made out "a long line of infantry, cavalry, and artillery, regularly drawn upon the plains of Argaum, immediately in front of that village."

"Although late in the day," says Wellesley in describing the events of the 29th November, "I immediately determined to attack this army. Accordingly, I marched on in one column, the British cavalry leading in a direction nearly parallel to that of the enemy's line;59 covering the rear and left by the Mogul and Mysore cavalry. The enemy's infantry and guns were on the left of their center, with a body of cavalry on their left. Sindhia's army, consisting of one massive body of cavalry, was on the right, having to its right a body of pindarries and other minor corps. Their line is extended above five miles, having in their rear the village and extensive gardens and enclosure of Argaum; and in their front a dry, which, however, was much cut by watercourses, etc.

"I formed the army in two lines; the infantry in the first, the cavalry in the second, and supporting the right; and the Mogul and Mysore cavalry the left, nearly parallel to that of the enemy; with the right rather advanced to press upon the enemy's left. Some little time elapsed before the lines could be formed, owing to a part of the infantry of my division which led the column having got into some confusion. When molded, the whole advanced in the greatest order; the 74th and 78th regiments were attacked by a large body (supposed to be Persians), and all these were destroyed. Sindhia's cavalry charged the 1st Battalion, 6th regiment, which was on the left of our line, and were repulsed; and their whole line retired in disorder before our troops, leaving in our hands 38 pieces of cannon and all their ammunition.

"The British cavalry then pursued them for several miles, destroyed great numbers, and took many elephants and camels and much baggage. The Mogul and Mysore cavalry also pursued the fugitives and did them great mischief. Some of the latter are still following them; and I have sent out this morning all of the Mysore, Mogul, and Marhattá cavalry, to secure as many advantages from this victory as can be gained, and complete the enemy's confusion.... The troops conducted themselves with their usual bravery...."

One of the bravest deeds performed during the battle of Argaum was that of Lieutenant Langlands, of the 74th. Wounded in the fleshy part of the leg by a spear, he promptly pulled out the weapon and thrust it through the body of the Arab who had thrown it. A Sepoy who witnessed this extraordinary display of self-possession, forgetting all discipline, rushed from the ranks and patted the young officer on the back, yelling in his native tongue, "Well done, sir; very well done!"

Mountain Fort of Gawilghur

Wellesley next marched on the mountain fort of Gawilghur, strongly garrisoned by the Rájá's troops. This defense consisted of an outer and inner fort, the former protected by strongly-built walls, and the whole by ramparts and towers. Admittance was gained only by three gates, all hard-hitting of access by an invading army owing to the roads leading to them. That to the south, communicating with the inner fort, was long and steep, and could only be negotiated on foot; the second was exposed to the guns mounted on the west side and was extremely narrow and scraped by rock; the third, or north gate, communicated with the village. Wellesley chose the last as being the most practicable for his purpose, although he did not blind his eyes to the fact that "the difficulty and labor of moving ordnance and stores from Labada would be very unlimited."

From the 7th December, when the corps under Wellesley and Stevenson marched from Ellichpúr by different routes, till the 12th, "on which Colonel Stevenson broke ground near Labada, the troops in his division went through a series of active services, such as I never before witnessed, with the utmost cheerfulness and perseverance. The heavy ordnance and stores were dragged by hand over mountains, and through ravines, for nearly the whole distance, by roads which it had been previously necessary for the troops to make for themselves."

On the night of the 12th, Stevenson erected two61 batteries in front of the north face of the fort, and Wellesley one on the mountain, "under the southern gate." Although firing was begun on the following morning, the breaches in the walls of the outer fort were not sufficiently large for practical purposes until the 14th. Next day, while the storming party was getting to work, Wellesley made two attacks from the southward so as to draw the enemy's fire upon himself as much as possible. The north-west gate was carried, and a detachment entered without difficulty. Captain Campbell, with the light infantry of the 94th, then succeeded in fixing ladders against the wall of the inner fort. They "escalated the wall, opened the gate for the storming party, and the fort was shortly in our possession." In a subsequent communication, Wellesley mentions that he never knew a place was taken by storm which was so little plundered. "And it is but doing justice to the corps to declare that in an hour after having stormed that large place, they marched out with as much regularity as if they had been only passing through it."

Bhonsla Rájá had already sent his vakeel [18] to sue for peace. This was granted by his ceding to the Company the province of Cuttack, with the district of Balasore, and dismissing the European officers who had played so important a part in the drilling of his army. Sindhia also "began to be a little alarmed respecting his own situation," and shortly afterwards concluded hostilities, handing over all the country between the Jumna and the Ganges, and several important fortresses. These happenings did not relieve Wellesley from active service. Several bands of freebooters, "the terror of the country," consisting mainly of fugitive soldiers from the defeated armies, were carrying on bad practices in the West Deccan. After crossing the Godavery, he and some of his troops marched many weary miles along bad roads, often at accelerated speed, to attack them, only to find that the enemy62 had received intelligence of their approach, probably from a traitor in Wellesley's own ranks. With definite purpose, the General continued to follow where the marauders led, and eventually broke up the bands, securing the whole of their guns, ammunition, and baggage, thus depriving them of their means of warfare: "they have lost everything which could enable them to subsist when collected." Wellesley afterwards asserted that his chase of the freebooters was the greatest march he ever made.

Breakup of the Army in the Deccan

Towards the end of May 1804, Wellesley received instructions from the Governor-General to break up the army in the Deccan, the task of running to earth Holkar, the sole remaining enemy of the Confederacy, being given to Lake. In the following month, he relinquished his command, and after a short visit to Calcutta returned to Seringapatam. He had already requested that he might be allowed to leave India "when circumstances permit it," and the Commander-in-Chief had given him the necessary permission. He was dissatisfied because he had not been promoted since he became Major-General, "and I think that there appears a prospect of service in Europe, in which I should be more likely to get forward." Also, he was suffering from rheumatism, "for which living in a tent during another monsoon is not an excellent remedy." He sailed for the Homeland on the 10th March 1805, after six years of hard work, and still harder fighting, in the interests of British rule in India.

The following contemporary pen-portrait of "the Sepoy General," sketched for us by Captain Sherer, will enable us to visualize him as he appeared at this time:

"General Wellesley was a little above the middle height, well limbed, and muscular; with the minimum encumbrance of flesh beyond that which gives shape and manliness to the outline of the figure. With a firm tread, an erect carriage, a countenance strongly patrician, both in the feature, profile, and expression. With an appearance remarkable and distinguished: few could approach him on any duty or on any subject requiring his serious attention, without being sensible of a something strange and penetrating in his perfect light eye. Nothing could be more simple and straightforward than the matter of what he uttered; nor did he ever in his life affect any peculiarity or pomp of manner, or rise to any course, weak loudness in his tone of voice. It was not so that he gave expression to excited feeling."

East India Company

To what extent did the Governor-General influence his brother's career in India? First of all, we must understand the position of Marquis Wellesley. It was naturally one of tremendous power and responsibility. The glamour attached to the post was sufficiently evident to the general public. There it ended, for it was glitter rather than gold to its holder. The Directors of the East India Company, ever on the side of the rigid economy and large dividends, expressly forbade the costly system of conquest and annexation, yet this was necessarily the sheet anchor of Wellesley's policy, as earlier chapters have shown. When pacific measures were tried and failed, it would have been disastrous to continue them. As it usually took over three months [19] for communication from India to reach England, it follows that the same period was necessary for a reply. The consequences of indecision on the part of the Viceroy, of waiting for advice from home in matters requiring urgency, were, therefore, fraught with dire peril. On the other hand, if he showed too autocratic tendencies he ran a grave risk of incurring displeasure. Indeed, this is correctly what happened, for Lord Wellesley was recalled in 1805 and censured by the Court of Proprietors. When, after thirty years, it became evident that his administration had been wise and not foolish, that he had carried out what would have had to be done eventually to establish British influence, the Directors relented and voted him a grant of £20,000.

Fortunately, there was "a barrier state" in London between the Governor-General and the Directors in the person of the President of the Board of Control, the said Board consisting of Cabinet Ministers. This position had been occupied since July 1802 by Lord Castlereagh, who, on taking office, found that Wellesley had come to the conclusion that resignation was better than humiliation. He did much to smooth over the difficulties, and from that time until Wellesley's return to England Castlereagh loyally supported the Viceroy on every possible occasion. For instance, when the reduction of the Indian establishment to 10,000 troops was earnestly mooted by the Directors and the Cabinet at home, notwithstanding the threatening attitude of the Marhattá Confederacy, it was principally due to Castlereagh's support of Lord Wellesley's demands that so absurd a policy was prevented.

The President of the Board of Control never interfered in the matter of patronage, knowing full well that the Governor-General on the spot was better able to recognize merit for

Duke of Wellington

the special requirements of the service than a man thousands of miles away. This brings us back to our proper subject.

We have noted how Lord Mornington discerned the opportunity awaiting his brother in India, and how that brother reciprocated when the former was diffident in the matter of accepting the key official post there. It is true that Wellesley was made Governor of Seringapatam over the head of Baird, his senior officer, but whether this appointment was because Mornington influenced General Harris in the matter of his choice is not sufficiently evident. There is a[65] strong suspicion that it was, [20] because Arthur Wellesley had only served as commander of the reserve, whereas Baird was the leader of the assault, and as such military tradition unquestionably favored his appointment.

Again, in the matter of the Batavian expedition, the Governor-General offered Wellesley, the appointment as a military commander. "The King has given me the power of selecting the persons who are to conduct this expedition; ... and a strict sense of duty induces me to think that you are the fittest person to be nominated for that service, provided you can safely be spared from Mysore for the period of the expedition...." In Mornington's opinion, "the expedition will be very advantageous to the naval and military commanders."

Duke of Wellington

On the other hand, we know that when the project was abandoned for a diversion on the coasts of the Red Sea, he superseded his brother. One wonders what would have happened when Wellesley set off for Bombay without instructions, had he not been closely related to the Governor-General. The Marquis certainly did not minimize Arthur's successes to those at home. Writing to Addington, then Speaker of the House of Commons, in October 1800, he says, "My brother Arthur has distinguished himself most brilliantly in an expedition against an insurgent, who had collected a great force of predatory cavalry—the wreck of Tippú's army." Three years later, when Addington was Prime Minister, he again drew attention to his brother's achievements, as follows: —

"My public duty will not permit me to be silent respecting Major-General Wellesley. His march from Mysore to Poona, his responsible conduct of the measures adopted for restoring the Peishwah. For conciliating the feudatory Mahratta chiefs who maintained their allegiance to the Peishwah. For preserving the dominions of the Nizám, and our interests at Hyderabad, combined with his sieges of Ahmednuggur, Burrampur, and Asseerghur. His glorious and splendid victories at Assaye and on the plains of Argaum, with the entire ruin of Sindhia's French troops and powerful artillery in the Deccan, must place the name of General

Duke of Wellington

Wellesley among the most bright and distinguished characters that have adorned the military history of the British power in India. He is now employed in reducing the central fortress of Perar, and in negotiating, with the ultimate judgment and skill, the conditions of peace. I leave his merits to your justice, and to the judgment of his King and country. The pride and honour of being allied by the nearest ties of blood to such an officer cannot absolve me from the obligations of my public station. As the representative of the supreme civil and military authority in India; and I cannot, therefore, omit this testimony to the merits of General Wellesley without an encouraging violation of my duty." [21]

Whatever may be thought of such glowing praise from a brother on the score of good taste, it evidently achieved its purpose, for before he left India, Arthur Wellesley was appointed a new Knight Companion of the Bath and received the thanks of the King and Parliament.

Earl Roberts, in summing up this phase of the future Duke's career, remarks: "On his arrival in India he found himself in a country where in almost every matter the power and influence of the Governor-General were supreme, and the Governor-General being his brother. He was quickly placed in a position of responsibility, which gave him the opportunity of developing his talents as a soldier and statesman in the best of all schools—the school of practice. It cannot be denied that in his early life Wellington owed much to family influence, [23] and to a system of promotion which would now be stigmatized as jobbery. On the other hand, he took full advantage of every chance that was thrown in his way, and by his industry and capacity fully justified the unique favor with which he was treated."

With this conclusion the present writer heartily agrees; whatever Sir Arthur gained from his relative's assistance was amply repaid in his achievements. British India owes much to the brothers Wellesley.

Duke of Wellington

CHAPTER VI
England, Ireland, and Denmark (1805-7)

"I am not afraid of responsibility, God knows, and I am ready to incur any personal risk for the public service."

Wellington.

Outline of specific point of views or inside knowledge in this chapter:

24. Shortly after his return from India Wellesley had his only interview with Nelson, an account of which is given in the "The Story of Nelson," pp. 113–4.
25. See his wife, p. 7.
26. "Personal Reminiscences of the first Duke of Wellington" (Edinburgh 1904), p. 274.
27. Lord-Lieutenant of Ireland.
28. At Copenhagen.
29. Flat-bottomed boats, usually armed with small guns.
30. Sir Herbert Maxwell, vol. i. p. 87.
31. Wilson is wrong in some of his facts. The Danish troops numbered some 14,000, and 1100 prisoners were taken. See Sir Herbert Maxwell, vol. i. p. 87.
32. "The Croker Papers," vol. ii. pp. 120–21.

33. H. W. Wilson, B.A., in "Cambridge Modern History," vol. ix. p. 236.
34. "The Life of Napoleon I," vol. ii. p. 143.

When, in 1803, the short-lived Peace of Amiens came to an end, and Great Britain and France again resorted to the sword, Napoleon's first feat of arms was the conquest of Hanover. Thus, at the very beginning of the second phase of the Great War, George III found himself not only minus his hereditary continental possessions but deprived of a very useful base for those futile military excursions so beloved of the British Government.

That His Majesty received the tidings of his loss "with great magnanimity, and a real kindliness of mind," may or may not be true. His ministers asserted that such was the case; considerations of policy would have precluded them from saying otherwise.

Command to Hanover

However, this may be, two months after Sir Arthur Wellesley landed in England, that is to say, in November 1805, he was given the command of a brigade in an expedition to Hanover about to be undertaken by Lord Cathcart. The object was to rout the comparatively few French troops left to garrison the country, and to co-operate with Russian, Swedish, and Danish troops in ridding Germany of the common enemy. The surrender of Mack at Ulm, and Napoleon's brilliant victory at Austerlitz, although it followed within a few weeks of Nelson's signal triumph at Trafalgar, [24] completely shattered this desirable object, just as the negotiations that followed put an end to the ambitious hopes of the Third Coalition. The recall of the troops before they had been able to carry out any of the objects of the diversion, beyond gaining some thousands of adherents to the rank and file, therefore, became imperative and was duly effected.

Sir Arthur Wellesley now spent a short time in command of his brigade at Hastings, and he was gazette colonel of the famous 33rd Regiment, which post had become vacant on the death of the venerable Marquis Cornwallis, his brother's successor in India. The next important event in his life, if not in his career, was his marriage to the Hon. Catherine Pakenham, thus consummating a romance began many years before, [25] and his personal ambition apart from the Army. The ceremony was performed in Dublin on the 10th April 1806, the bridegroom being nearly thirty-seven years of age. One wishes it was possible to add that "they lived happily ever after." Biography, the twin sister of History, tells us that it was not so, and Gleig suggests that a broken engagement with a second suitor, of which Wellesley was not informed on his return from India, was partly the cause.

Two days after the wedding Wellesley was elected Member of Parliament for Rye, his primary object in seeking political distinction being that he might defend his brother's administration in India, where his system of making recalcitrant States subsidiary to England while retaining their own rulers, was the subject of an embittered attack. The "high crimes and misdemeanours" alleged against Lord Wellesley were referred to from time to time. On the 17th March 1808, the following motion was carried by 182 votes against 31: "That it appears to this House that Marquis Wellesley, in his arrangements in the province of Oude, was actuated by an ardent zeal for the service of his country. With an anxious desire to promote the safety, interests, and prosperity of the British Empire in India." This did not altogether end the unsavory affair, for another unsuccessful attempt to incriminate the statesman was made sometime later.

Chief Secretary for Ireland

Sir Arthur was by this time Chief Secretary for Ireland, having been appointed in the previous year. Once again we see two members of this distinguished family holding prominent appointments, for Henry Wellesley became one of the

Duke of Wellington

Secretaries to the Treasury in the newly-appointed Portland ministry.

Barrington, whose acquaintance we have already made, relates an attention-grabbing anecdote of the soldier at this time. He met Lord Castlereagh, accompanied by a gentleman, in the Strand. "His Lordship stopped me," he writes, "whereas I was rather surprised, as we had not met for some time; he spoke very kindly, smiled, and asked if I had forgotten my old friend, Sir Arthur Wellesley? Whom I discovered in his companion, but looking so sallow and wan, and with every mark of what is called a worn-out man, that I was strictly concerned with his appearance. But he soon recovered his health and looks, and went as the Duke of Richmond's secretary to Ireland, where he was in all material traits still Sir Arthur Wellesley, but it was Sir Arthur Wellesley judiciously improved. He had not forgotten his friends, nor did he forget himself. He said that he had accepted the office of secretary only on the terms that it should not impede or interfere with his military pursuits, and what he said proved true...."

Understandably his duties in Ireland bear no comparison with those he so successfully undertook in India, but following his own maxim, "to do the business of the day in the day," he got through a vast amount of general labor, frequently chief, sometimes trivial. Under the former head, we must put his investigation of the military defenses of the island. It must not be forgotten that although the invasion of the United Kingdom by Napoleon was no longer a standing menace, there was always a likelihood of its resurrection, and Ireland was the danger zone.

Peace of Tilsit

The Peace of Tilsit signed between France and Russia on the 7th July 1807, and between France and Prussia on the 9th of the same month, was a most shocking blow to British interests. By a secret treaty, Emperor Alexander undertook to aid Napoleon against England if that Power refused to make peace within a definite period, to recognize the equality of all nations at sea, and to hand back the conquests made by her since 1805. As a bait—it really savored of insult—Great Britain was to be offered Hanover. Should she refuse these terms, the Autocrats of France and of Russia agreed to compel Denmark, Sweden, and Portugal to join them in a vast naval confederacy against Great Britain, and to close their ports against her. Also, the reigning monarchs of Spain and Portugal were to be deposed in favour of the Bonaparte family. For his connivance in the matter, Alexander was to be handsomely compensated in the Ottoman Empire and by territorial acquisitions in Western Europe.

Duke of Wellington

Fortunately, or otherwise, according to the point of view, the British Cabinet was put in possession of certain facts regarding these plans. Canning, who was Minister for Foreign Affairs, realizing the responsibilities of his unenviable position, as also of that of his country, determined to forestall the plotters. He felt that some kind of arrangement with Denmark was essential, especially as the Prince Regent of Portugal had communicated news to the effect that Napoleon purposed to invade England with the Portuguese and Danish fleets. Canning suggested to Denmark that her fleet should be put in the safe custody of England until peace was restored. Also, he promised a subsidy of £100,000, and the assistance of troops should Denmark be attacked. Mr. F. J. Jackson was sent to open negotiations; the Prince Royal promptly vetoed them. "I stated plainly," says Jackson, "that I was ordered to demand the junction of the Danish fleet with that of England, and that in the case of refusal it was the determination of His Majesty to enforce it."

Lord Cathcart was put in command of an army of 27,000 troops, the naval portion of the expedition being placed in the hands of Admiral Gambier. No sooner had Sir Arthur Wellesley heard of the project than he communicated with Castlereagh, then at the War Office and ever his staunch supporter, for an opportunity to take part. He was given charge of a division. On the 3rd August, a formidable array of twenty-five sail-of-the-line and over fifty gunboats and transports appeared off Elsinore. Gambier and Cathcart were told by Jackson "that it now rested with them to carry out the measure prescribed by the British Government." In a letter to his brother the diplomatist adds, "The Danes must, I think, soon surrender, for they are without any hopes of succour. They are unfurnished with any effectual means of resistance, and are almost in total want of the necessaries of life, as far as I could learn or was able to see for myself during my few hours' stay there. **28** There were no droves of cattle or flocks of sheep; no provisions of any sort being sent in the direction of the city. No troops marching towards the town; no guns mounted on the ramparts; no embrasures cut, in fact, no preparations of any sort. What the Danes chiefly rely on is the defense by water. They brought out this morning several *praams* **29** and floating batteries, and cut away one or two of the buoys.

"The garrison of Copenhagen does not amount to more than four thousand regular troops. The *Landwehr* is a mere rabble, as indeed all *levées en masse* must be.

"The people are said to be anxious to capitulate before a conflagration takes place, which must happen soon after a bombardment begins, when, not improbably, the fleet, as well as the city, will become a prey to the flames."

Jackson's prophecy came true, but against his statement that the army disembarked at Veldbeck "in grand style," we must set that of Captain Napier: "I never saw any fair in Ireland so confused as the landing. If the enemy had opposed us, the *remains* of the army would have been on their way to England." **30** Wellesley's first affray—it can scarcely be termed a battle—took place at Roskilde. Like almost everything connected with the expedition, Jackson has something to say about it, and that "something" in this particular instance is anything but complimentary. "Sir Arthur Wellesley," he tells his wife, "has had an affair which you will probably see blazoned forth in an extraordinary *Gazette*. With about four thousand men, he attacked a Danish corps of the armed peasantry, and killed and wounded about nine hundred men, besides taking upwards of fifteen hundred prisoners, amongst whom were sixty officers. One was a General officer. I spoke to him this morning, for him and his officers are let off on their parole. The men are on board prison ships, and miserable wretches they are, fit for nothing but following the plough. They wear red and green striped woolen jackets, and wooden *sabots*. Their long, lank hair hangs over their shoulders and gives to their rugged features a natural expression. The wise ones say, that after the first fire, they threw away their arms and hoping, without them, to escape the pursuit of our troops. In fact, the *battle* was not a very glorious one, but this you will keep for yourself." **31**

Wellesley himself afterwards referred to the event as "the little battle at Kiöge," and mentioned that "the Danes had made but a poor resistance; indeed, I believe they were only new raised men—militia." **32**

The bombardment of Copenhagen

Began on the 2nd September 1807, and concluded three days later, when an armistice was granted so that terms might be discussed. On the 7th, Copenhagen capitulated. The conditions imposed by Sir Arthur Wellesley, Sir Home Popham and Lieutenant-Colonel Murray were that the British should occupy the citadel and dockyards for six weeks, and take possession of the ships and naval stores. Their troops would then evacuate Zealand. "I might have carried our terms higher ... had not our troops been needed at home," Wellesley writes to Canning. The various clauses were agreed out, and fifteen sail-of-the-line, fifteen frigates, and thirty-one smaller vessels of the Danish fleet, as well as 20,000 tons of naval stores, were escorted to England. "That the attack was necessary," says a recent historian, "no one will now deny. England was fighting for her existence; and, however, disagreeable was the task of striking a weak neutral, she risked her own safety if she left in Napoleon's hand a fleet of such proportions. In Count Vandal's words, she 'merely broke, before he had seized it, the weapon which Napoleon had determined to make his own.'" **33** Dr J. Holland Rose disapproves, and points out, that "In one respect our action was unpardonable: it was not the last desperate effort of a long period of struggle. It came after a time of selfish torpor fatal alike to our reputation and the interests of our allies. After protesting their inability to help them, Ministers belied their own words by the energy with which they acted against a small State." **34**

Canning's hope for an alliance with Sweden, to keep open the Baltic, was destined never to be fulfilled. Sir John Moore was sent to assist Gustavus in his efforts to resist the attacks of Russia, but the nation deserted the King, deposed him, and joined Napoleon. War speedily broke out between Sweden and Denmark, and also between the Denmark and Great Britain. The Czar's overtures to England on behalf of France, as arranged at Tilsit, came to nothing. He was not anxious for them to have any other ending, so enraptured was he with Napoleon's grandiloquent schemes. Enraptured? Yes, but only for a few short years.

CHAPTER VII
Battles of the Peninsular War (1808)

"In war men are nothing: it is a man who is everything."

Napoleon.

Outline of specific point of views or inside knowledge in this chapter:

35. See Oman's "Peninsular War," vol i. pp. 1–11.
36. Oman, vol. i. pp. 631–639. Returns of October-November 1808.
37. Succeeded by Soult in November 1808.
38. Oman, vol. i. pp. 640–45.
39. "The Autobiography of Sir Harry Smith," 1787–1819. Edited by G. C. Moore Smith, M.A. (London Ed. 1910).
40. "A Boy in the Peninsular War," edited by Julian Sturgis (London, 1899), p. 313.
41. *Ibid.* p. 311.
42. Vol. i. p. 235 n.
43. The total loss of the regiment was 190, by far the heaviest of those engaged.

Portugal & Spain

On his return from Copenhagen, Wellesley, never happy unless his mind was fully occupied, resumed his duties as Chief Secretary for Ireland. Special mention of the services he had rendered to his country was made in the House of Commons, and there was some talk of a second period in India, where affairs were far from settled. Before long, however, it became increasingly evident that his knowledge and ability would be required nearer home.

WELLINGTON'S PENINSULAR CAMPAIGNS.

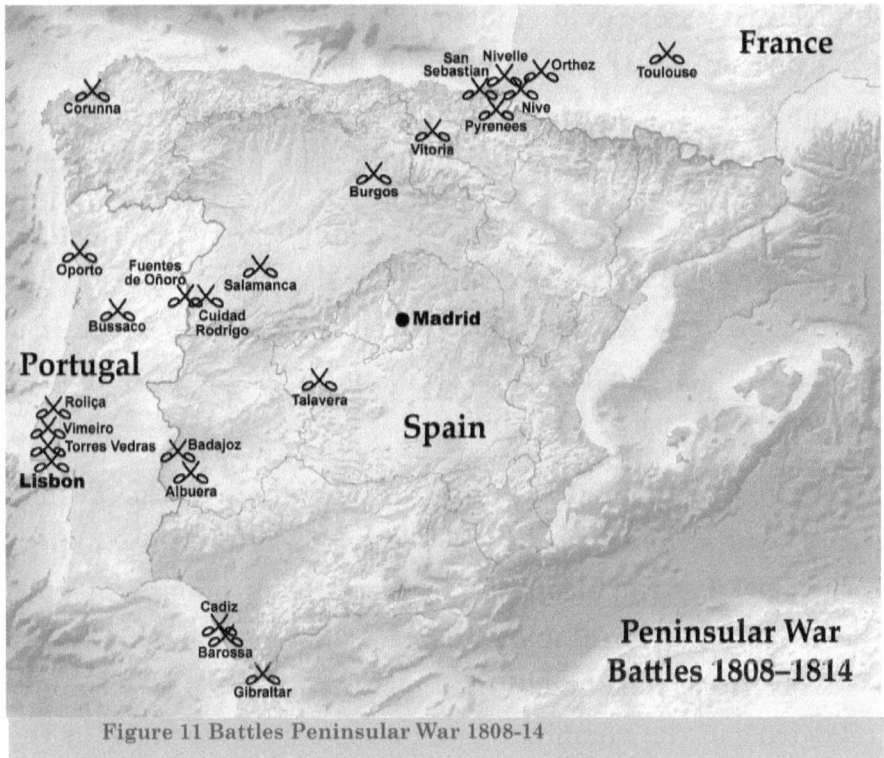

Figure 11 Battles Peninsular War 1808-14

Portugal, our old ally, had been forced by Napoleon to declare war against Great Britain on the 20th October 1807. Bent on pursuing the rigid restrictions on trade imposed by his Continental System, he had also peremptorily ordered the confiscation of the property of the British merchants. Fortunately for those most concerned, the Prince Regent remembered past friendship and may have discerned future possibilities. He temporized, and this enabled many of the

Duke of Wellington

English residents to settle their affairs and sail for home before the Dictator could enforce obedience. The sequel was the overrunning of the kingdom by French troops under the intrepid Junot, who met with no resistance, and the desertion of their subjects by the Royal Family, who sailed for Brazil.

Although this plan was carried out at the earnest request of the British Government, as represented by Lord Strangford, the Ambassador at the Portuguese Capital, it cannot be regarded as a suitable example of patriotism on the part of the House of Braganza.

In October 1807, Junot, in command of the French Army, and strengthened by a few regiments of the Spanish corps placed at Napoleon's disposal for the dismemberment of the western portion of the Iberian Peninsula, began his march on Lisbon. He concluded it on the 30th November with only 1500 troops, the remainder following slowly because of the terrible sufferings they had endured during a forced march made at Napoleon's urgent behest.

Here it should be mentioned that the presence of the Spanish troops was due to the infamous Treaty of Fontainebleau, signed the previous October. In this arrangement, the Emperor had promised Godoy, the real ruler of Spain and an intensely ambitious man, a large slice of territory in the country about to be conquered in return for favors rendered.

Duke of Wellington

It is more than probable that Napoleon never intended this particular clause to be taken seriously by anyone but his dupe; the gift was so much dust thrown in the eyes of the favorite for the purpose of securing the entry of French troops into Spain. **35** In this, he was pre-eminently successful. Once in Lisbon Junot speedily removed any fear of the national army by breaking up many of the regiments and sending the remainder on service outside the kingdom. The flames of rebellion were not yet kindled. So far so good.

Unhappily the grand prizes which the Emperor had hoped to secure at Lisbon were beyond his reach.79 Even the squadron which was to have seized the Portuguese and British shipping in the harbor was held in check by the hated English.

Napoleon, pretending to be the friend of Spain, was in reality her worst enemy. He merely used her as a useful tool to pick Portuguese locks, and then pursued the same course with his friend's lockers. He began his unwelcome attentions by seizing the important frontier fortresses of Pampeluna, Barcelona, San Sebastian, and Figueras, and invading the country by a force which speedily numbered 116,000 men, mostly conscripts, for he thought the country easy prey. Murat entered Madrid as Junot had entered Lisbon. By the most questionable methods, namely, the enforced abdication of

Charles IV and his son Ferdinand, the Emperor secured the throne, permanently as he fondly imagined, for his brother Joseph, King of Naples.

In July 1808, the eldest Bonaparte was proclaimed King and entered his capital. Within a month he found it desirable to retire behind the Ebro; his subjects had not only broken into open revolt, but a French army of over 17,000 troops under Dupont had been forced to capitulate at Baylen, in Andalusia. Riots, assassinations, and massacres made it evident that the Spanish temper was considerably more dangerous than that of the Portuguese; it soon became apparent, moreover, that the people had employed some of their time in organizing, on a necessarily rough and ready principle, such forces as they possessed.

The inhabitants of the Asturias, in the north, were the first of the provincials to apply the torch to the tinder of revolt, after a riot in Madrid on the 2nd May 1808, and its Junta General called into being a levy of 18,000 men to protect the principality. It sent two deputies to England for assistance, which was readily given in money and military stores. Other provinces likewise selected Juntas, and Galicia also dispatched representatives80 to plead its cause in London. Galicia, adjoining the Asturias on the west, lost little time in following the warlike example of its neighbors, and the arsenals of Coruña and Ferrol, made memorable by the Trafalgar campaign, threw in their lot against Napoleon and contributed no fewer than thirty-two battalions of regulars and militia to the regular forces. Leon and Old Castile also rose in rebellion, though with less energy. There were too many French in the Basque Provinces and Navarre for much to be attempted there. Coming still farther to the east, Catalonia sheltered 16,000 regulars and many different levies, but Aragon, Valencia, and Murcia were fragile. Andalusia, in the extreme south of the country, was almost as fortunately placed about troops as Galicia, and the remains of the French fleet which had escaped Nelson and Collingwood were taken as they rode in Cadiz harbor.

Nothing was approaching to go into action, provinces and towns often viewing in more or less friendly rivalry. They did not understand, or if they assumed they did not realize, that patriotic cliques do not make for strength. They fought for themselves rather than for the nation as a whole. Throughout the struggle, we find a lack of cohesion.

When we come to look at the earliest available statistics 36 of the various Spanish armies which formed the front line, we find that their total strength in regulars, militia battalions, and newly raised corps was 151,248. They were divided into five main armies, namely, of Galicia, Aragon, Estremadura, the Centre, and Catalonia, under Generals Blake, Palafox, Galluzzo, Castaños, and Vives respectively. The troops of the second line numbered about 65,000 and included the Army of Granada, under Reding, the Army of Reserve of Madrid, commanded by San Juan, the Galician, Asturian, Estremaduran, Andalusian, Murcian and81 Valencian reserves, and the 3000 odd men in garrison in the Balearic Isles.

The gross total of the French Army of Spain at this period dwarfs the above figures for all their brave show; it reached 314,612. From this must be deducted 32,643 detached troops and 37,844 in a hospital or missing, making the "operational" no fewer than 244,125. Of the eight corps, Victor commanded the 1st, Bessières **37** the 2nd, Moncey the 3rd, Lefebvre the 4th, Mortier the 5th, Ney the 6th, St Cyr the 7th, and Junot the 8th. There were also Reserve Cavalry and Infantry, the Imperial Guard, troops marching from Germany, and National Guards on the French frontier. **38**

When we consider that on the 31st May 1808 Napoleon had only 116,000 men in Spain and that within six months he had found it necessary to more than double that number, the critical nature of the undertaking becomes modest.

Duke of Wellington

To enter fully into the doings of the various armies throughout the war would deflect us far out of our proper course, but we shall hear of them whenever Wellesley was involved.

If you would know the fierce spirit of the patriots, the hate they cherished for Napoleon and the French, you have only to turn to any one of the many Memoirs of men who fought in the Peninsular War. Captain, later Sir Harry, Smith, who was with Sir John Moore in 1808 and remained with Wellesley until March 1814, gives many instances in his vivacious "Autobiography," 39 but the following must suffice. Smith's guide happened to be the owner of the house in which his wife and baggage were quartered in the village of Offala:

82 "After I had dressed myself," he relates, "he came to me and said, 'When you feast, I have some capital wine, as much as you and your servants like; but,' he says, 'come down and look at my cellar.' The fellow had been so civil; I did not like to refuse him. We descended by a stone staircase, he carrying a light. He had upon his countenance a most sinister expression. I saw something exceedingly excited him: his look became fiend-like. He and I were alone, but such confidence had we Englishmen in a Spaniard, and with the best reason, that I apprehended no own evil. Still his appearance was very singular. When we got to the cellar-door, he opened it and

held the light so as to show the cellar. When, in a voice of thunder, and with an expression of demoniacal hatred and antipathy, pointing to the floor, he exclaimed, 'There lie four of the Devils who thought to subjugate Spain! I am a Navarrese. I was born free from all foreign invasion, and this right hand shall plunge this stiletto in my own heart as it did into theirs, ere my countrymen and I are subjugated!' brandishing his weapon like a demon. I see the excited patriot as I write. Horror-struck as I was, the instinct of self-preservation induced me to admire the deed exceedingly, while my simple frame quivered and my blood was frozen, to see the noble science of war and the honor and chivalry of arms reduced to the practices of midnight assassins. Upon the expression of my admiration, he cooled, and while he was deliberately drawing wine for my dinner, which, however, strange it may be, I drank with the gusto its flavor merited, I examined the four bodies. They were Dragoons—four strong, healthy-looking fellows. As we ascended, he had seamlessly recovered the equilibrium of his vivacity and naturally good humor. I asked him how he, single-handed, had perpetrated this deed on four armed men (for their swords were by their sides). 'Oh, merely enough. I pretended to love a Frenchman'[83] (or, in his words, 'I was an Afrancesado'), 'and I proposed, after giving them a good dinner, we should drink to the extermination of the English.' He then looked at me and ground his teeth. 'The French rascals, they little guessed what I contemplated. Well, we got into the cellar, and drank away until I made them so drunk, they fell, and my purpose was easily, and as joyfully, effected.' He again brandished his dagger, and said, 'Thus die all enemies to Spain.' Their horses were in his stable. When the French Regiment marched off, he gave these to some guerrillas in the neighborhood. It is not difficult to reconcile with truth the assertion of the historian who puts down the loss of the French army. During the Spanish war, as 400,000 men, for more men fell in this midnight manner than by the broad-day sword, or the pestilence of climate, which in Spain, in the autumn, is excessive."

Duke of Wellington

Figure 12 Sir Harry Smith

Duke of Wellington

That there was significant cause for complaint on the part of the Spaniards is also borne out by other eye-witnesses. Napier records that a captain and his company came across a peasant's hut and demanded provisions, as was their accustomed. The father explained that his children were half-starving, and he had, but the little food left. He was told that he would be hanged to a beam. Should he give a sign that he repented of his decision he would be cut down, but not otherwise? He was strung up without further ado. Then the cries of his wife and children overcame his noble act of self-sacrifice, and he was released. The soldiers then took every scrap of food in the miserable dwelling and departed. A similar method was adopted by a second body of plunderers, and when they could find nothing they spitefully killed the poor fellow, doubtless on the charge that he was hiding his stock.

Robert Blakeney, in noticing that most writers have referred to the Spanish army as "ragged, half-famished wretches," cautions us that the men themselves must not be blamed for their unkempt appearance.84 "The scandal and disgrace," he writes, "were the proper attributes of the Spanish Government. The members of the Cortez and Juntas were entirely occupied in speculation, amassing wealth for themselves and appointing their relatives and dependents to all places of power and emolument. Even if unworthy and unqualified. Although it was notorious that shiploads of arms, equipment, clothing and millions of dollars were sent from England. Sent for the use and maintenance of the Spanish troops, yet all was appropriated to themselves by the members of the general or local governments or their rapacious satellites while their armies were left barefoot, ragged and half-starved. In this deplorable state they were brought into the field under leaders, many of whom were scarcely competent to command a sergeant's outlying picquet. For in the Spanish army, as elsewhere, such was the undue influence of a jealous and covetous aristocracy, that, unsupported by their influence, personal gallantry and distinction, however conspicuous, were but rarely rewarded."

40 The same officer, who joined the 28th Regiment as a boy of fifteen and saw much service in the Peninsular War, assures us that "Courage never wanted to the Spanish soldiers, but confidence in their chiefs was rare." **41**

An expedition against the American colonies of Spain had been mooted several times by the British Cabinet, and Sir Arthur Wellesley had reported on ways and means. The scheme had developed sufficiently for some 8000 troops to be assembled at Cork preparatory to embarking for the voyage. It was finally decided that the troops should be used for a descent on Portugal, with the immediate intention of expelling

Duke of Wellington

the French and raising the enthusiasm of the population in contradiction of Napoleon.

The force sailed on the 12th July 1808 with Wellesley, now a Lieutenant-General, in command.

John Wilson Croker, who served his country as Secretary to the Admiralty from 1809 to 1830, dined with Sir Arthur and Lady Wellesley in Harley Street on the evening before the General set out for Cork. After settling some business connected with Ireland, Wellesley "seemed to lapse into a kind of reverie," his guest informs us, "and remained silent so long that I asked him what he was thinking of. He replied, 'Why, to say the truth, I am thinking of the French that I am going to fight. I have not seen them since the campaign in Flanders when they were great soldiers, and a dozen years of victory under Buonaparte must have made them better still. They have besides, it seems, a new system of strategy, which has out-maneuvered and overwhelmed all the armies of Europe. 'Tis enough to make one thoughtful; but no matter: my die is cast, they may overwhelm me, but I don't think they will out-maneuver me. First, because I am not afraid of them, as everybody else seems to be; and secondly, because if what I hear of their system of maneuvers be true, I think it a false one as against loyal troops. I suspect all the continental armies were more than half beaten before the battle was begun. I, at least, will not be frightened beforehand."

Wellesley in Portugal

Wellesley made the voyage to Portugal in a fast frigate, and landed at Coruña on the 20th July 1808, ahead of his troops. This gave him sufficient time to do a preliminary study of the situation at first hand, and to be ready for immediate operations on the arrival of his men.

The first news he received was not encouraging, for it told of the battle of Medina de Rio Seco, which Bessières had won against the Army of Galicia on the 14th July. A little relief was afforded by rumors of success elsewhere, and "the arrival of the British money," speedily renewed the flagging spirits of the patriots who were fighting under such adverse conditions.

The Junta of Galicia, while keenly appreciative of gold, ammunition, and arms, showed no disposition to avail themselves of the Commander's services, and suggested his landing in the north of Portugal as the government of Oporto was collecting native troops in that neighborhood. "The difference between any two men," Wellesley writes on the 21st July, the day before he sailed from Coruña, "is whether the one is a better or a worse Spaniard, and the better Spaniard is the one who detests the French most heartily. I understand that there is actually no French party in the country, and at all events I am convinced that no man now dares to show that he is a friend to the French."

To sum up the situation was not an arduous task for Wellesley. He came to the conclusion without further ado that the only reasonable way to assist the Spaniards was "to get possession of and organize a regular army in Portugal." He proceeded to the fleet off Cape Finisterre, spent a few hours there, and then went to Oporto, where he had an important conference with the Bishop, who was also head of the Portuguese Junta, and some military officers. It was eventually decided that about 5300 troops, chiefly infantry, stationed at Coimbra under Bernardino Freire. These troups should be used to co-operate with Wellesley, and that the remaining forces, namely, 12,000 peasants, should either be employed in the neighbourhood or in the province of Tras os Montes, where a French attack seemed probable. Finally, a spot in Montego Bay was chosen as the most suitable point for disembarkation, especially as it had the additional advantage of being near Coimbra. On the 1st August the business commenced, tiresome, and not unattended by danger because of the heavy surf.

Duke of Wellington

Wellesley had much to think about while this was proceeding. He had just received the amazing news[87] that he had been superseded by Sir Hew Dalrymple, with Sir Harry Burrard as second in command, that Sir John Moore was on his way with 10,000 men, and that he (Wellesley) and Lieut.-Generals the Hon. J. Hope, Sir E. Paget, and Mackenzie Frazer were to command divisions. Whatever agitation the new arrangements may have occasioned Wellesley; he did not allow it to shake his purpose or lessen his enthusiasm for the cause he had now so much at heart. He writes to Castlereagh, "Whether I am to command the army or not or am to quit it, I shall do my best to ensure its success. You may depend upon it that I shall not hurry the operations, or commence them one moment sooner than they ought to be initiated so that I may acquire the credit of the success. The Government will determine for me what way they will employ me hereafter, whether here or elsewhere." He then goes on to sketch a campaign suitable for an army "of 30,000 Portuguese troops, which might be easily raised at an early period; and 20,000 British, including 4000 or 5000 cavalries."

"The weather was so rough and stormy," writes one of the soldiers of the 71st Regiment, "that we were not all landed until the 5th. On our leaving the ship, each man got four pounds of biscuit, and four pounds of salt beef cooked on board. We marched, for twelve miles, up to the knees in the sand, which caused us to suffer much from thirst; for the marching made it rise and cover us. We lost four men of our regiment, who died of thirst. We buried them where they fell. At night, we came to our campground [Lugar], in a wood, where we found plenty of water, to us more acceptable than anything besides on earth. We here built large huts and remained four days. We again commenced our march amongst the coast, towards Lisbon. In our advance, we found all the villages deserted, except by the old and destitute...."

On the night of the 8th, General Spencer and his corps of 4500 men joined Wellesley from Cadiz, where he had landed at the request of the Junta of Seville. By the 11th, the whole army had arrived at Leiria, and on the following day it was augmented by 2300 of Freire's Portuguese troops, their commander refusing point blank to march with his remaining forces unless certain impossible demands were met. "My object," writes Wellesley, "is to obtain possession of Lisbon, and to that I must adhere, whatever may be the consequences, till I shall have attained it, as being the first and greatest step towards dispossessing the French of Portugal."

Meanwhile, Junot had sent instructions to Generals Loison and Delaborde to effect a junction and attack Wellesley. This was prevented by the timely arrival of the British

Duke of Wellington

troops at Leiria, for the former was some sixteen miles to the south-east and the latter about the same distance to the south-west. Wellesley was consequent among them. This necessitated Loison's return to the southward if he wished to join Delaborde, and the British General determined to prevent the operation. On the 14th, Wellesley was at Alcobaço, from whence the French had retreated but a few hours before.

Although a small engagement took place near Obidos, Wellesley did not offer battle until two days later because his whole force had not yet come up. The conflict occurred at Roliça, where Delaborde's army was awaiting him on a hill. We know that the allied force totaled 15,000; the strength of the enemy is uncertain, Wellesley believing it to be 6000 while Professor Oman [42] gives the figure as "about 4350 men," basing his conclusion on known official returns previous to the fight and making allowance for probable losses by sickness.

"On the morning of the 17th," says the eye-witness already quoted, "we were under arms an hour before day. Half an hour after sunrise, we observed the enemy in a wood. We received orders to retreat. Having fallen back about two miles, we struck to the right, to come upon their flank, during the 9th, 29th, and 5th battalion of the 60th attacked them in front. They had a solid position on a hill. The 29th advanced up the hill, not perceiving an ambush of the enemy, which they had placed on each side of the road. As soon as the 29th was right between them, they gave a volley, which killed, or wounded, every man in the grenadier company, except seven. Unmindful of their loss, the regiment drove on and carried the entrenchments. **43** The engagement lasted until about four o'clock when the enemy gave way. We continued the pursuit, till darkness put a stop to it. The 71st had only one man killed and one wounded. We were maneuvering all day, to turn their flank; so that our fatigue was excessive though our loss was small."

Such was the battle of Roliça, Wellesley's first victory over the French. He was perfectly satisfied with the fighting and moral qualities of his men as displayed in this engagement.

"I cannot sufficiently applaud the conduct of the troops throughout this action," he tells Castlereagh. Although he had a superiority of strength, the number of soldiers "actually employed in the heat of the action," namely, 4635, was, "from unfavorable circumstances ... by no means equal to that of the enemy." The returns showed 479 British killed, wounded, and missing, and the French about 600.

Duke of Wellington

CHAPTER VIII
Victory Abroad, Anger at Home (1808-9)

"From the sublime to the ridiculous is but a step."

Napoleon.

Outline of specific point of views or inside knowledge in this chapter:

44. The case of Peter Findlater at Dargai is almost an exact parallel.
45. See also some remarks in "The Croker Papers," vol. ii. pp. 121–22.

With a mere handful of soldiers, Junot, big with ideas of a future kingship, and underestimating the strength and fighting powers of the enemy, left Lisbon and entered the field against Wellesley, whose troops were now encamped at Vimiero to cover the landing of 4000 other men under Generals Anstruther and Acland. Having joined forces with the unfortunate Loison and Delaborde and thereby brought up the total strength of his army to 13,056 men, the Marshal prepared to attack.

Duke of Wellington

Wellesley, who had over 18,000 troops, including 2000 Portuguese, was well prepared, nay eager, for the encounter, but, unfortunately for him, Burrard arrived on the evening of the 20th August. When Wellesley explained to him his scheme of operations, he showed no disposition to fall in with it. Wellesley had wished Sir John Moore to proceed to Lisbon by land to cut off Junot's retreat, but the less-active Burrard would have none of it, and ordered him to wait until Moore's arrival. "Whether we advance or not," replied the General, "we shall have to fight.91 For the French will certainly attack us if we do not attack them."

Duke of Wellington

This prophecy was fulfilled about 8 o'clock on the morning of the 21st August 1808 when squadrons of the enemy's cavalry appeared. An attack was made on the British advanced guard. The French were driven back at the point of the bayonet, while other troops, stationed in the churchyard

Figure 13 This miniature was painted 1808 at the outset of the Peninsular War

of Vimiero, prevented them from reaching the village of that name, and Acland's brigade attacked them in the flank. "A most desperate contest" was necessary before the enemy recoiled in confusion, during which they lost heavily in killed and wounded, and in *material* seven pieces of cannon. Other French troops, supported by a large body of cavalry, turned their attention to the heights on the road to Lourinhão, where Ferguson's brigade was stationed. The latter charged with praiseworthy coolness, and again there was a tale of disaster to tell when the enemy fell back while half a dozen guns were captured. An attempt to recover part of the lost artillery resulted in the French being obliged to retire "with great loss."

Burrard, who had slept on the vessel which had brought him out, did not arrive on the field till late in the day, and took no part in the direction of the battle until Wellesley wished to pursue the enemy to Torres Vedras and cut them off from Lisbon. "Sir Harry," he said, "now is your time to advance, the enemy is utterly beaten, and we shall be in Lisbon in three days." This his senior officer absolutely forbade. Had the former been allowed to follow his own wishes he believed that, "in all probability, the whole would have been destroyed." As it was, at least 1800 of the enemy were rendered *hors de combat*, including 300 or 400 troops who were made prisoners. The British lost in killed and missing 186 men, and 534 were wounded. The General was again delighted with the behavior of his men, and in communicating with the Duke of92 York, he averred that "this is the only action I have ever been in, in which everything passed as it was directed and no mistake was made by any of the Officers charged with its conduct."

One rare incident, one altogether human touch, affords relief to the story of the battle of Vimiero. A piper of the gallant 71st Highlanders, severely wounded in the thigh and deeply in need of surgical aid, continued to blow his pibroch for the encouragement of his colleagues until exhaustion finally conquered his determined spirit. Seated on the ground, he declared that "the lads should not want music to their walk," and went on with his weird music as though parading within the walls of Edinburgh Castle.

"I afterwards saw him," relates Lieut.-General Sir William Warre, "in a hovel, where we collected the wounded ... both French and English. I shook him by the hand and told him I was very sorry to see so fine a fellow so badly hurt. He answered, 'Indeed, Captain, I fear I am done for, but there are some of those poor fellows,' pointing to the French, 'who are terrible indeed.'"

Such coolness, typified in successive instances, although not always under such conditions, has made our Empire what it is to-day. The "common" British soldier, sowing the highway with his bones, enables a later generation to reap a golden harvest.

It is due to the French to record that they were not without men equally as cool as Piper Mackay. A typical example is furnished by Major Ross-Lewin, who fought in the 32nd, and it occurred immediately after the battle of Vimiero:

"An officer of my regiment," he relates, "happened to pass near an old French soldier, who was seated by the roadside, covered with dust, and desperately wounded. A cannon-shot had taken off both his feet just above the ankles, but his legs were so swollen that his wounds bled but little. On seeing the officer, the poor fellow addressed him, saying, '*Monsieur, je vous conjure donnez moi mes pieds.*' and at the same time pointed to his feet, which lay on the road beyond his reach. His request met with a ready compliance. The pale, toilworn features of the veteran brightened up for an instant on receiving these mutilated members, which had borne him through many a weary day, and which it grieved him to see trampled on by the victorious troops that passed. Afterwards, as if prepared to meet his fast-approaching fate becomingly, by the attainment of this one poor wish, he laid them tranquilly beside him, and, with a look of resignation, and the words, '*Je suis content,*' seemed to settle himself for death."

Duke of Wellington

Many years afterwards, when in a reminiscent mood, the Duke of Wellington recapitulated the events of the 21st August 1808. "The French," he told his guests, "came on at Vimiero with more confidence, and seemed to *feel their way* less than [smiling] I always found them to do *afterwards*. They came on in their usual way, in a large column, and I received them in line, which they were not accustomed to, and we repulsed them there several times. At last they went off beaten on all points while I had half the army untouched and ready to pursue. However Sir H. Burrard—who had joined the army in about the middle of the battle, but seeing all doing so well, had desired me to continue in the command now that he considered the battle as won though I thought it half done—resolved to push it no further. I begged very hard that he would go on, but he said enough had been done. Indeed, if he had come earlier, the battle would not have taken place at all. That is when I waited on him on board the frigate in the Bay the evening before, he desired me to suspend all operations and said he would do nothing till he had collected all the force which he knew to be on the way. He had heard of Moore's arrival, but the French luckily resolving to attack us, led to a different result. I came from the frigate about nine at night and went to my own quarters with the army, which, from the nearness of the enemy, I naturally kept on the alert. In the dead of the night a fellow came in—a German sergeant, or quartermaster—in a great fright—so great that his hair seemed actual to stand on end—who told me that the enemy was advancing rapidly, and would be soon on us. I immediately sent round to the generals to order them to get the troops under arms, and soon after the dawn of day we were vigorously attacked. The enemy were first met by the (50th?), not a good-looking regiment, but devilish steady, who received them admirably, and brought them to a full stop immediately, and soon drove them back; they then tried two other attacks ... one severe, through a valley on our left; but they were defeated everywhere, and completely repulsed, and in full retreat by noon, so that we had time enough to have *finished them* if I could have persuaded Sir H. Burrard to go on."

Battle of Vimiero

Battle of Vimiero, 21 August 1808

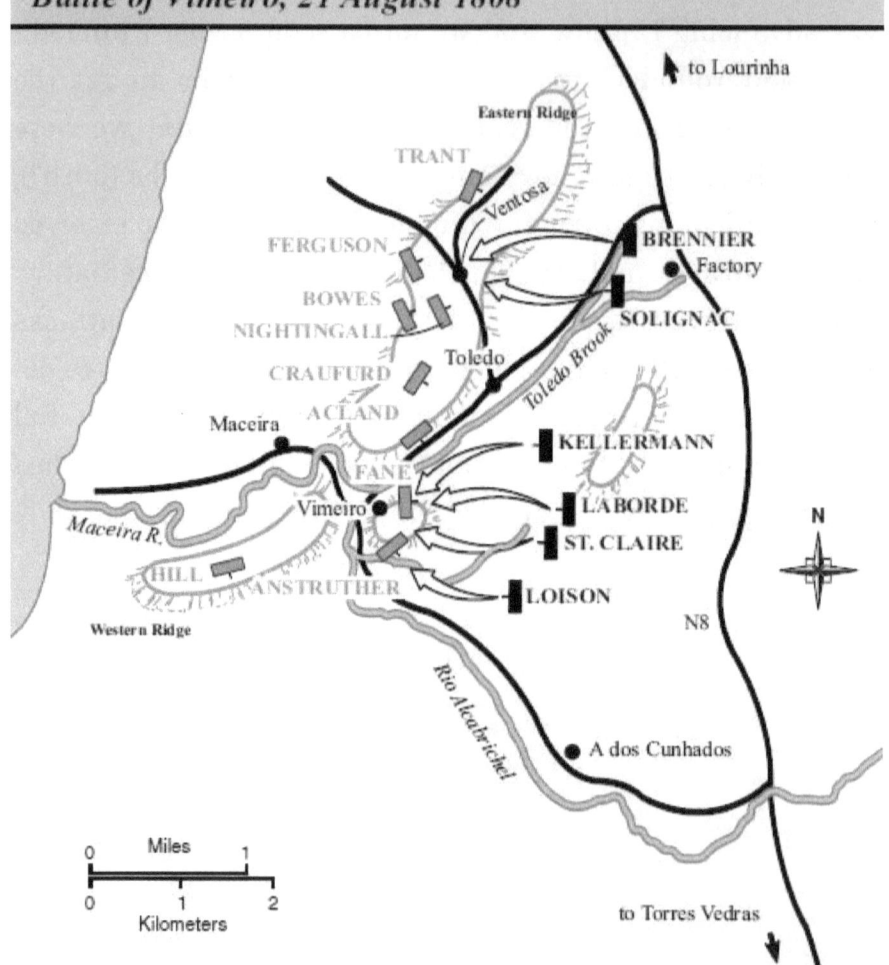

Figure 14 Battle of Vimiero - Positions

Duke of Wellington

On the day following the battle of Vimiero, Dalrymple arrived. While pondering over the situation he received a proposal for an armistice from Junot, which developed into the Convention of Cintra, preliminarily signed on the 30th August 1808. The most rudimentary conditions were—the surrender of all places and forts in Portugal occupied by the French troops, the evacuation of the country, and the transport of the army, its munitions and "property," to France in British ships. By a strange oversight, the important question of future service was overlooked. Consequently, there was nothing to prevent an early return of the troops to the Peninsula should Napoleon think fit for them to do so.

We have now to consider Wellesley's part in this much-discussed transaction. The Convention was definitely signed on the 30th August 1808, but previous to this a meeting of the General Officers was called to deliberate upon it. "The result of the meeting," Wellesley writes on the 29th inst., "was a proposal to make certain alterations, which I acknowledge I do not think sufficient, although the treaty will answer in its amended form.... At the same time, I must say that I approve of allowing the French to evacuate Portugal because I see clearly that we cannot get them out of Portugal. Under existing circumstances, without such an arrangement; and we should be employed in the blockade or siege of the places which they would occupy during the season in which we ought and might be advantageously engaged against the French in Spain. But the Convention, by which they should be allowed to evacuate Portugal, ought to be settled in the most honorable manner by the army by which they have been beaten. We ought not to be kept for 10 days on our field of battle before the enemy (who sued on the day after the action) is brought to terms.

"I am quite annoyed on this subject."

Wellesley signed the preliminary Memorandum at the request of Dalrymple but had nothing to do with the final settlement. "I lament the situation of our affairs as much as you do," he writes on the 5th September, "and I did everything in my power to prevent it; but my opinion was overruled. I had nothing to do with the Convention as it now stands, and I have never seen it to this moment.... I have only to regret that I put my name to an agreement of which I did not approve, and which I did not negotiate. If I had not done it, I really believe that they would not have dared to make such a Convention as they have made: notwithstanding that that agreement was never ratified, and is now so much waste paper." 45

His letters at this period teem with allusions to the unfortunate treaty. He tells Castlereagh that "It is quite impossible for me to continue any longer with this army. I wish, therefore, that you would allow me to return home and resume the duties of my office. If I should still be in office, and it is convenient to the Government that I should retain it. If not, that I should remain upon the Staff in England. If that should not be practicable, that I should persist without employment. You will hear from others of the various causes which I must have for being dissatisfied, not only with the military and other public measures of the Commander-in-Chief but with his treatment of myself. I am convinced it is better for him, for the army, and for me, that I should go away; and the sooner I go, the better."

On the 6th, October Wellesley was in London, and at once resumed his office as Chief Secretary for Ireland. The newspapers teemed with unsavory references to the unpopular Convention; the caricaturists, not to be rivalled by their journalistic brethren, produced the grossest lampoons for the benefit of the indignant public. In one of them Wellesley and his colleagues are hanging on gibbets, in another the former is shown urging his troops to glory:

This is Sir Arthur (whose valor and skill, began so well, but ended so ill) Who beat the French, who took the Gold, that lay in the City of Lisbon.

Duke of Wellington

Windham, writing in his Diary under date of the 16th September, probably sums up the thoughts of most British statesmen of the time: "At Chesterford heard the report of news; said to be excellent, but without particulars. Feasted upon the hopes of what I should meet at Hockrill. Alas! *Quanti de spe decidi!* It was the news of the convention with Junot. *There never was surely such a proceeding in the history of wars or negotiations.* There is no bearing the thought of it."

A Court of Inquiry was instituted. Dalrymple and Burrard were recalled, and together with Wellesley, were examined by a board of officers, which included General David Dundas and Lord Moira, at Chelsea Hospital. The finding of the Court was non-committal "respecting the fitness of the Convention in the relative situation of the two armies," doubtless because a unanimous "verdict" could not be arrived at, but the members undeniably declared "that unquestionable zeal and firmness appear throughout to have been exhibited by Lieut.-Generals Sir Hew Dalrymple, Sir Harry Burrard, and Sir Arthur Wellesley...." In commenting on the judgment thus expressed, Sir Herbert Maxwell notes that the two senior officers were never employed again, adding, "Similar eclipse might have fallen upon Sir Arthur, but for the efforts of Castlereagh and other influential friends, whose confidence in their General was never shaken."

Duke of Wellington

House of Lords

In the following January (1809) the House of Lords and the House of Commons expressed their thanks to General Wellesley for the victories of Roliça and Vimiero.

"It is your praise," said the Speaker of the Commons, "to have inspired your troops with unshaken confidence and unbounded ardour. You have commanded, not the obedience alone, but the hearts and affections of your companions in arms. Having planned your operations with the skill and promptitude which have so eminently characterized all your former exertions, you have again led the armies of your country to battle, with the same deliberate valour and triumphant success which have long since rendered your name illustrious in the remotest parts of this Empire.

"Military glory has ever been dear to this nation; and great military exploits, in the field or upon the ocean, have their real reward in Royal favor and the gratitude of Parliament. It is, therefore, with the highest satisfaction, that, in this recent instance, I now proceed to deliver to you the thanks of this House...."

Wellesley's reply was made in three well-chosen98 sentences, without the slightest attempt at rhetoric. In the House

Duke of Wellington

of Lords, Vimiero was spoken of as "a signal victory, honorable and glorious to the British arms." The resolutions of the peers, which included high appreciation of the behavior of the non-commissioned officers and privates, were conveyed to Sir Arthur by Lord Chancellor, and acknowledged by their recipient in a short letter, the most key paragraphs of which are as follows:

"I have received the mark of distinction which the House of Lords have conferred upon me with sentiments of gratitude and respect proportionate to the high sense. I entertain of the greatness of the honor which it carries with it, and I shall have great pleasure in communicating to the Officers and the troops the special reward of their exemplary conduct which their Lordships have conferred upon them.

"I beg leave, at the same time, to express to their Lordships my thanks for the expressions of personal civility with which your Lordship has conveyed to me the commands of the House."

These signs of approval must have been entirely satisfactory to Sir Arthur after the bitter criticisms of the previous months, but what he particularly valued was a handsome service of plate, worth intrinsically £1000. However sentimentally beyond price, presented to him by the Brigadier and field officers who were associated with him in the victory at Vimiero. They, at any rate, had implicit faith in their General.

Duke of Wellington

CHAPTER IX
Sir Arthur's Return to Portugal (1809)

"We are not naturally a military people; the whole business of an army upon service is foreign to our habits, and is a constraint upon them, particularly in a poor country like this."

Wellington.

Outline of specific point of views or inside knowledge in this chapter:

46. As to the merits and demerits of national resistance, see some thoughtful remarks in Arnold's "Introductory Lectures on Modern History," pp. 158–64.
47. See also some pointed remarks in Wellesley's dispatch dated Badajoz, 21st November 1809. It will be remembered that at the time of the Russian-Japanese war, newspaper men were wisely precluded from publishing particulars of proposed movements and same intelligence likely to be of service to the enemy. During the recent conflict between Italy and Turkey, the most rigid censorship was exercised by the former Power.
48. "I rather think that Mortier had removed from Zaragoza, but some time elapsed before he arrived in Old Castile."—Note by Wellesley.

49. "The Adventures of a Soldier," by Edward Costello.
50. Oman, vol. ii. p. 334. This disposes of the often-repeated story that Waters discovered the little craft in the reeds. Brailmont, for instance, says that the Colonel "suddenly darted off from the throng," and half an hour later the skiff "shot out into the deep" with six men on board.

Baron de Frénilly, travelling to Paris in December 1808, notes that "the roads along which we passed were crowded with splendid troops who were on their way to find a grave in the Peninsula." Napoleon, in the Constitution he granted to Spain, assumes for himself not only the so-called "divine right of kings," but the special favor of Providence. "God," he says, "has given me the power and the will to overcome all obstacles." Frénilly, writing after the Emperor's death, merely states an historical fact; Napoleon, at the height of his stupendous power, regards himself as omnipotent, and proves within a few years that he is not.

Yet it must be conceded that the Dictator of Europe—apart from moral considerations, which never troubled him to any extent—had a certain right to infer from his past experience that the Almighty was on his side. It was not for him to foresee that the Peninsula was to prove a running sore of the Imperial body politic. To be sure, Joseph had not been particularly successful on the throne of the Spanish Bourbons. Murat had displayed many silly qualities, Dupont had surrendered, Junot had evacuated Portugal, and eleven million had rebelled either practically or theoretically against French domination, but there was still himself, and God was "on the side of the heaviest battalions!" "I may find in Spain the Pillars of Hercules, but not the limits of my power." Thus he endeavored to encourage his brother, and there is no reason to suspect that he imagined otherwise. He announced that he would pour between 300,000 and 400,000 troops across the Pyrenees. He actually began the new campaign with over 200,000, which compared more than favourably with the 120,000 ill-organized Patriots under Castaños, Palafox, Vives, Belvedere, Blake, and La Romana, who usually acted without any idea of the value of co-operation.

The number of those ready and willing to engage in a guerilla warfare cannot be given. 46 Statistics fails in such a matter like this. Names indelibly associated with Napoleon's greatness were either present or coming—Victor, Bessières, Moncey, Lefebvre, Ney, St Cyr, Mortier, and Junot.

When Dalrymple, Burrard, and Wellesley sailed from Portugal, the British command devolved upon Sir John Moore. This being a biography of Wellington, Moore's astounding campaign can only be referred to in the briefest way, but it is necessary to mention the most important incidents if we are to understand the various phases of the war. Leaving 9000 men at Lisbon with Lieutenant-General Sir John Cradock, and taking with him 14,000 troops, Moore advanced into Spain to co-operate with the Spaniards according to his instructions. His own columns reached Salamanca, the point of concentration, in November 1808, but Baird, who, with a reinforcement of 13,000 men, was to effect a junction with him, found it impossible to do so. There was much delay in consequence.

In the first week of the following month, the Emperor was at Madrid, and the Spanish capital once again in the hands of the French. Disaster after disaster had followed hard in the tracks of the national forces.

It was Moore's hope that by slowly retreating northward the enemy would follow, and thus enable his allies in the south to recover. Having united with Baird, and learned that Soult, with not more than 20,000, was near Sahagun, Moore was on the eve of combat when the startling intelligence reached him that Napoleon was in pursuit. The Emperor had told the Senate, "I am determined to carry on the war with the whole activity, and to destroy the armies that England has disembarked in that country." With wonderful promptitude, Moore turned towards Coruña, where he believed the British fleet awaited him. Napoleon, hearing disconcerting news from Paris, made off for his capital, leaving Soult, "the Iron Duke of France," and Ney to pursue the red-coats.

On the 16th January 1809, the battle in which Moore received his death-wound was fought. Within twenty-four hours, the victorious troops embarked for the homeland. Not a British soldier, other than deserters or stragglers, was left in Spain. In the sister kingdom there were some 12,000, of whom 9000 had been left at Lisbon by Moore when he had set out for Salamanca; the remainder had arrived from England in the early November and December. Also, Sir Robert Wilson had succeeded in equipping some 1300 men at Oporto for his Loyal Lusitanian Legion.

It soon became evident that war would shortly break out between France and Austria, thus precluding any thought on Napoleon's part of going back to the Peninsula. Castlereagh, notwithstanding previous experiences, was as enthusiastic as ever. Baird and his ragged troops were no sooner home than it was remarked that the Secretary for War and Wellesley were very frequently together. Wiseacres foretold an early return of the latter to Spain and Portugal.

Duke of Wellington

Memorandum of Portugal

Sir Arthur prepared a lengthy Memorandum on the Defence of Portugal, which was placed in the hands of the Cabinet for careful consideration. "I have always been of opinion," it begins, "that Portugal might be defended, whatever might be the result of the contest in Spain; and that in the interim the measures adopted for the defense of Portugal would be highly useful to the Spaniards in their fight with the French."

Wellesley suggested the thorough reorganization of the native Portuguese troops, part of the expense being borne by Great Britain, and the employment of not less than 30,000 British troops, including 4000 or 5000 cavalries and a large body of artillery. The entire army was to be commanded by British officers. Riflemen and 3000 British or German cavalry should be sent as additional reinforcements as soon as possible, in addition to a corps of engineers for an army of 60,000. He perfectly understood that the French would not be caught napping, for "it may be depended upon that as soon as the newspapers shall have announced the departure of Officers for Portugal. The French armies in Spain will receive orders to make their movements towards Portugal, so as to anticipate our measures for its defence. We ought, therefore, to have everything on the spot, or nearly so before any alarm is created at home respecting our intentions."

Wellesley in Portugal

Thanks in no small measure to Castlereagh, Wellesley was appointed to the supreme command of the new 103 expedition. He left England on the 14th April 1809, a few weeks after Soult's Vanguard had crossed the Portuguese frontier, and landed at Lisbon on the 22nd, after a most eventful voyage, having encountered terrible weather off the Isle of Wight, which threatened to drive his vessel ashore. The Commander-in-Chief thus sums up the situation on the Peninsula: "At that time," he says, "the French had got possession of Zaragoza, Marshal Soult held Oporto and the northern provinces of Portugal. The battle of Medellin had been fought on the 29th March; and General Cuesta was endeavoring to recover from its effects, and to collect an army again at Monasterio, in the mountains of the Sierra Morena. The French, under Marshal Victor, were in possession of the Guadiana and had their excellent posts as forward as Los Santos. Sebastiani was at Ciudad Real, and held in check the army of La Carolina, at that time under the command of General Venegas, consisting of about 12,000 men. Ney was in possession of Galicia; Salamanca was held by a small detachment of French troops; St

Cyr was at Catalonia with his corps of 25,000 men; and Kellermann, who had succeeded to Bessières in the command of the 6th corps, was at Valladolid. Mortier, with his corps, and the Duc d'Abrantès (Junot), with the 8th corps, at Zaragoza. The Portuguese army was totally disorganized and nearly annihilated, and the Spanish troops were scarcely able to hold their positions in the Sierra Morena. The Marquis de la Romana, who had been with his corps on the frontiers of Portugal, near Chaves, from the period of the embarkation of the British army at Coruña, in the month of January, until the month of March, had moved from thence when Soult invaded Portugal by Chaves. Afterwards moved towards the Asturias with his army, and went himself into that province."

104 The greeting the Commander-in-Chief received at the hands of the populace of Lisbon would have been embarrassing to one possessing a less cool head, but Wellesley knew perfectly well that applause to-day is apt to become condemnation to-morrow. He was appointed Marshal-General in the Portuguese Army, which was now placed in the capable hands of General Beresford by the British Cabinet, Wellesley's one second-in-command being Major-General Rowland Hill. According to his instructions, "the defense of Portugal you will consider as the first and immediate object of your attention. But, as the security of Portugal can only be effectually

Duke of Wellington

provided for in connection with the defence of the Peninsula in the larger sense. His Majesty on this account, as well as from the unabated interest he takes in the cause of Spain. He leaves it to your judgment to decide, when your army shall be advanced on the frontier of Portugal, how your efforts can be best combined with the Spanish, as well as the Portuguese troops, in support of the common cause. In any movements you may undertake, you will, however, keep in mind that, until you receive further orders, your operations must necessarily be conducted with especial reference to the protection of that country."

Of British troops the Commander-in-Chief now had at his disposal 23,455, namely, 18,935 infantries, 4270 cavalries, and 250 attached to the wagon train; Portugal contributed 16,000 men. Costello, a non-commissioned officer of the 95th Rifles, and later a Captain in the British Legion, has nothing good to say of the Portuguese troops. In his record of the Peninsular War [49] he gives several instances of their unreliability and treacherous nature. One example must suffice:

"The sanguinary nature of the Portuguese," he says, "during the whole period of the war was notorious. When crossed or excited, nothing but the shedding of [105] blood could allay their passion. It was always with the greatest difficulty

Duke of Wellington

that we could preserve our French prisoners from being butchered by them, even in cold blood. They would hang upon the rear of a detachment with detainees, like so many carrion birds, waiting for every opportunity to satiate their love of vengeance, and it required all the firmness and vigilance of our troops to keep them in check. It was well known that even our men fell in stepping between them and the French whom they had marked out as victims. Indeed, it was not infrequent for our men to suffer from the consequences of their ferocity, and I, while at Vallée, had a narrow escape. I had crossed the hills to purchase some necessaries at the quarters of the 52nd Regiment, and on my return fell in with several of the soldiers of the 3rd Caçadores. One of them, a fierce-looking scoundrel, evinced a great inclination to quarrel, the more particularly as he perceived that I was unarmed and alone. Having replied rather sharply to some abuse they had cast upon the English, by reflecting on their countrymen in return, he flew into a rage, drew his bayonet, and made a rush at me, which I avoided by stepping aside, and tripping him head foremost on the ground. I was in the act of seizing his bayonet, when some his comrades came up, to whom he related, in abstract terms, the cause of our disagreement. Before he had half concluded, a general cry arose of 'Kill the English dog'; and the whole, drawing their bayonets, were advancing upon me when a party of the 52nd came up, the tables were turned, and the Caçadores fled in all directions."

Wellesley at once prepared to advance and had not been at Lisbon a week before representations were made to him by the Junta of Spanish Estremadura for aid in behalf of the southern provinces. He replied that until "the enemy who has invaded Portugal shall have been removed" he could not hope to lend them 106 the requisite assistance. "The enemy" consisted of the corps under Soult and Victor. The army of the former, which had left Coruña and invaded Portugal by Napoleon's imperative orders, now occupied Oporto. This, the second city in the kingdom and the center of the most prosperous district, fell after a gallant if unscientific resistance on the part of the inhabitants under the Bishop. The Marshal took dire vengeance on the insurgents during the journey from Galicia, perhaps because he had suffered from the guerilla warfare, now almost universal throughout the Peninsula. Victor's army was to take Badajoz and afterwards Seville while Sebastiani held the south in check. Wellesley decided to attack Soult, which necessitated a march of over eighty miles of rugged country. Leaving two small detachments of his own and of Portuguese troops upon the Tagus to watch the movements of Victor, he set out with 13,000 British, 9000 Portuguese, and 3000 Germans—25,000 in all.

Duke of Wellington

The right column, consisting of 9000 men under Beresford, was sent to Lamego, on the Douro, so as to be ready to cut off Soult's retreat, the left making for Oporto with Wellesley. On the 10th May, the cavalry and advanced guard of the latter crossed the Vouga, hoping to surprise the French troops at Albergaria and the neighboring villages, but the movement was not absolutely prosperous, although some prisoners and cannon were taken.

The advanced guard reached Vendas Novas on the following day and drove in the outposts of the French advanced guard. The latter were vigorously attacked in the woods and village and defeated with considerable loss. All this augured well for the following operations, and on the 12th the vanguard reached the southern bank of the Douro. The French were stationed on the opposite bank, having taken the precaution to burn the bridge after crossing, and to withdraw all the boats they could discover.

Unfortunate in this matter, Wellesley was favored in another. His army was screened by cliffs and a hill called the Serra. This bold rock was surmounted by a Franciscan convent, where the Commander posted batteries and made his observations. As the river winds a great deal, his movements were unobserved by those on the look-out at the French headquarters, to the left of Wellesley as he peered through his

Duke of Wellington

glass across the wide waterway. He had already perceived a huge building, known as the Seminary, surrounded by high walls with but one entrance on the landward side, and open to the river. This he knew would be an excellent position to secure, especially as it was almost opposite to him.

There was in the army a certain Colonel Waters, a keen-eyed officer with an infinite amount of resource and a ready wit. He contended that it was scarcely probable that Soult could have secured every boat, and interrogated a refugee on the point. He found that the man had crossed in a small skiff. 50 With the aid of the prior of Amarante, the fugitive, and several peasants, he heaved the boat out of the mud, and, crossing the stream, brought back some barges. In these three companies of the Buffs, under Lieut.-General Paget effected a landing on the opposite side. This outstanding officer was seriously wounded almost immediately afterward, but the passage of the Douro had been secured.

General Murray, who had been ordered to cross at Barca d'Avintas, also managed to get over, and signally failed to check the retiring columns after the battle. As additional troops gained the opposite shore, the French made repeated attempts to hurl108 them back but were ultimately obliged to retreat "in the utmost confusion" towards Amarante. According to a letter from General Stewart to his brother, Lord

Castlereagh, the French fled with such haste that "Sir Arthur Wellesley dined at their headquarters on the dinner which had been prepared for Marshal Soult."

On the 19th May, Soult was across the frontier, having been compelled to abandon over fifty guns and his baggage. In making his way across the Sierra Catalina to escape the pursuing troops, his rear-guard was defeated at Salamonde, with severe loss. He eventually reached Orense, in Galicia, minus some 5000 men, including the sick and wounded he had left behind him in Oporto.

"The road from Penafiel to Montealegre," says Wellesley, "is strewed with the carcasses of horses and mules, and of French soldiers, who were put to death by the peasantry before our advanced guard could save them. This last circumstance is the natural effect of the species of warfare which the enemy have carried on in this country. Their soldiers have plundered and murdered the peasantry at their pleasure. I have seen many persons hanging in the trees by the sides of the road, executed for no reason that I could learn, excepting that they have not been friendly to the French invasion and usurpation of the government of their country. The route of their column, on their retreat, could be traced by the smoke of the villages to which they set the fire."

Duke of Wellington

Within a fortnight, Wellesley was writing of the defects of his own men. "I have long been of opinion," he says, "that a British army could bear neither success nor failure, and I have had manifest proofs of the truth of this belief in the first of its branches in the recent conduct of the soldiers of this army. They have plundered the country most terribly, which has given me the greatest concern....

"They have robbed the people of bullocks, for what reason I am sure I do not know. Apart from, as I understand is their practice, to sell them to the people again. I shall be very much obliged to you if you will mention this practice to the Ministers of the Regency, and bid them issue a proclamation forbidding the people, in the most encouraging terms, to purchase anything from the soldiers of the British army."

The Commander-in-Chief relates the same disturbing facts to Castlereagh. "The army behaves terribly ill," is his expression. "They are a rabble who cannot bear success any more than Sir J. Moore's army could tolerate failure. I am endeavoring to tame them; but, if I should not succeed, I must make an official complaint about them, and send one or two corps home in disgrace. They plunder in all directions."

Meanwhile Victor, far from taking Badajoz and marching on Seville as the Emperor wished, had found it necessary to move on the route to Madrid, where he could secure much-needed assistance. He, therefore, took up his position at Talavera. Wellesley, intent upon crushing him, arrived at Abrantes about the same time as the Marshal was evacuating Estremadura and consequently abandoning the fruits of his victory over Cuesta at Medellin. For Wellesley to have followed Victor with the relatively few men at his disposal would have been to court disaster, and he, therefore, acquiesced in a new plan of operations suggested by Cuesta, in which the Spaniards were to be given an opportunity of showing their capabilities or proving their incapacity. This, says Professor Oman, was "the first and only campaign which he ever undertook in company with a Spanish colleague and without absolute control over the whole conduct of affairs."

CHAPTER X
Talavera (1809)

"The battle of Talavera was the hardest fought of modern times."

Wellington.

Outline of specific point of views or inside knowledge in this chapter:

51. At the approach of the enemy, no fewer than 6000 Spaniards took to their heels and played no part in the battle.
52. Napoleon made a similar error of judgment at Waterloo by keeping the Imperial Guard in reserve until after 7 p.m. (See *post*, p. 222).
53. Sir Herbert Maxwell, vol. i. p. 165, says 6268; Professor Oman ("Cambridge Modern History," vol. ix. p. 452) gives 5300, the Spanish casualties "trifling." The final authority states that 7200 Frenchmen were killed or wounded.

As to the
The potentialities of the new project were distinctly promising. After uniting with Cuesta, Wellesley was to follow the course of the Tagus and cut off Victor's army of 33,000

troops while the attention of Sebastiani and Joseph Bonaparte, who had but 17,000 men all, told, was occupied by Venegas.

When last heard of Soult and Ney were in Galicia busily engaged in suppressing an insurrection, so no opposition was anticipated from them. In this matter after events proved the facts to be far different from the surmise. There seemed every likelihood of a successful issue provided there was no snapping of individual links of the chain of operations. Wellesley did not find Cuesta, an above all pleasant colleague, but he was not the man to assert his own opinion unless he thought it imperative. He characterized him as having "no military genius," which is certainly more favorable than "that deformed-looking lump of pride, ignorance, and treachery," which is the description given to us by Costello. "He was," the latter adds, "the most murderous-looking old man I ever saw." They came together at Oropesa on the 20th July, their forces111 totaling 55,000, of which 35,000 were Spanish. It was the task of Venegas and his 26,000 men to approach Madrid from the south, and, by a demonstration in force, distract the attention of Sebastiani. He proved far too slow, and ere he was able to interfere, Victor, Sebastiani, and Joseph concentrated in the neighborhood of Toledo. On the 26th July, the last batch of their 50,000 troops came together.

Duke of Wellington

Had Wellesley been allowed to attack on the 23rd July, as he wished, it is probable that he would have crushed Victor, whose reinforcements did not begin to arrive until the following day. Cuesta had already shown his incompetence, and some of his advanced guards had been roughly handled by a French cavalry division. It was Wellesley's opinion that the psychological moment had arrived, but the Spanish commander objected. "Had we fought then," Wellesley afterwards averred, "it would have been as great a battle as Waterloo, and would have cleared Spain of the French for that time." The formidable task before him was not made lighter by the knowledge that the commissariat and transport arrangements had utterly broken down.

At Talavera, evacuated by Victor, who moved a few miles to the east, Wellesley was obliged to halt, and even threatened to withdraw from Spain because of the ill-treatment accorded his famishing troops. "I have never seen an army," he says, "so ill-treated in any country, or, considering that all depends upon its operations, one which deserved good treatment so much. It is ridiculous to pretend that the country cannot supply our wants. The French army is well fed, and the soldiers who are taken in good health, and well supplied with bread, of which indeed they left a small magazine behind them. This is a rich country in corn, in comparison with Portugal, and yet, during the whole of my operations in that country, we never wanted bread but on one day on the frontiers of Galicia. In the Vera de Plasencia, there are means to supply this army112 for 4 months, as I am informed, and yet the alcaldes have not performed their engagements with me. The Spanish army has plenty of everything, and we alone, upon whom everything depends on, are actually starving."

After considerable trouble, Cuesta consented to Wellesley assuming supreme command of the combined forces. On the afternoon of the 27th, the British General mounted his horse and, accompanied by his staff, rode out of the town to an old château, known as the Casa de Salinas. His object was to obtain a bird's-eye view from the roof of the movements of the enemy on the Alberche. He apprehended no danger because Spanish troops occupied the adjacent woods. In this, he was deceived, for some French *tirailleurs* suddenly appearing, the troops beat a hasty retreat. The Commander-in-Chief jumped from the wall and regained his horse not a moment too soon. Had it not been for the immediate presence of a body of English infantry, who immediately opened fire, it is extremely probable that Wellington and his staff would have been captured.

Duke of Wellington

At five o'clock the opposing forces were within touch, the French having crossed the river and driven in the British piquets, who lost about 400 men.

One of the finest descriptions of the ensuing battle—or more correctly, series of battles—is that of Captain M. de Rocca, a French officer of Hussars, which has the advantage of giving the point of view of the enemy, and how Wellesley was regarded by one at least of his combatants.

"The Spaniards," he says, "were posted in a situation deemed impregnable, behind old walls and garden fences, which border and encompass the city of Talavera. 51 Their right was defended by the Tagus, and their left joined the English, near a redoubt constructed on an eminence. The ground in front of the Anglo-Spanish armies was very unequal, and intersected113 here and there by ravines, formed by the rains of winter. The whole extent of their position was covered by the channel of an appealing deep torrent, at that time dry. The English left was strengthened by a conical eminence that commanded the greater part of the field of battle, and which was separated by a deep, extensive valley from the Castilian chain of mountains.

"This eminence was thus in a manner the key to the enemy's position, and against this decisive point of attack, an experienced general, possessed of that intuitive glance which ensures success, would immediately have led the principal part of his excessive force, to obtain possession of it. He would either have taken it by assault or have turned it by the valley. But King Joseph, when he should have acted, was seized with a broken spirit of indecision and uncertainty. He attempted only half measures, he distributed his forces partially and lost the opportunity of conquering while feeling the way for it. Marshal Jourdan, the second in command, had not that spur of patriotism in the Spanish war, which inspired him when he fought in the plains of Fleurus, to achieve the independence of France.

"The French commenced the engagement by a cannonade and rifle-fire in advance of their right, and they dispatched a single battalion only, and some sharp-shooters, by the valley, to take the eminence which defended the English left, never thinking they would do otherwise than yield. This battalion, however, having to contend with superior numbers, was repulsed with loss, and compelled to retire. A division of dragoons, which had gone to reconnoiter Talavera, found the approaches to that city strongly fortified with artillery, and could not advance.

"At nightfall, the French made another attempt to gain the hill. A regiment of infantry, followed at a short distance

Duke of Wellington

by two others, attacked the extreme left of the English with new ardor, arrived at the summit of the hill, and took possession of it. But having114 been fiercely assaulted, in its turn, by an entire division of the English, just, when having conquered, it was breathless with exertion, it was immediately obliged to give way. One of the two regiments, commanded to assist in this attack, had lost its way in a wood on account of the darkness; the other not getting soon enough over the ravine, which covered the enemy's position, had not arrived in time.

"Both these attacks had miscarried though conducted with heroic bravery because they had been made by an inadequate number of troops. A single battalion had been sent, and then one division when an enormous proportion of the whole army should have been dispatched. These unsuccessful attempts revealed to the English what we designed next day; and still more evidently demonstrated the importance of the station they held. They passed the greater part of the night in fortifying it with artillery.

"The sun rose next morning on the two armies drawn up in battle order, and again the cannonade commenced. The defense of Portugal being entrusted to the English army, the fate of that country, and, perhaps, of all the Peninsula, was now to be decided by this contest. The veterans of the first and

Duke of Wellington

fourth French corps, accustomed for years to conquer throughout Europe, and always to witness their ardor seconded by the combined skill of their chiefs, burned with impatience for orders to engage, and thought to overthrow all before them by one well-conjoined assault.

"One division alone, of three regiments of infantry, was sent to the valley to storm the position, of which we had, for a moment, obtained possession the preceding evening. After the considerable loss, this division reached the top of the eminence and was just about taking it. One of the regiments had already advanced as far as the artillery when their charge was repulsed, and the whole division was forced to retire. The English, apprehending by this renewed attack that the French designed to turn their left by the valley, stationed their cavalry there; and caused a division of the Spaniards to occupy the skirts of the high Castilian mountains beyond it. The French receded to the ground they first occupied. The cannonade continued for another hour and then became gradually silent. The intense heat of mid-day obliged both armies to suspend the combat, and observe a kind of involuntary truce, during which the wounded were removed.

"King Joseph, having at last gone himself to reconnoiter the enemy's position, gave orders, at four o'clock, for a general attack against the army of England. A division of dragoons was left to observe the Spaniards in the direction of Talavera. General Sebastiani's corps marched against the right of the English, while Marshal Victor's three divisions of infantry, followed by masses of cavalry, charged against their left, to attack the eminence by the valley. King Joseph and Marshal Jourdan took part with the reserve, in the rear of the 4th division. The artillery and musketry were not long in being heard.

English Commander stationed on the hill

Which overlooked the field of battle, was present always where danger demanded his presence. He could survey, at a glance, every corps of his army, and perceive below him the least movement of the French. He saw the line of battle formed, the columns disposed of for the conflict; he penetrated their designs by their arrangements, and thus had time to order his plans, so as to penetrate and prevent those of his foes. The position of the English army was naturally sturdy and difficult of approach, both in front and flank; but in the rear it was quite accessible, and gave ample freedom to their troops to hasten to the quarter threatened.

"The French had a ravine to pass before they could reach the enemy. They had to advance over ground much intersected, very rugged and unequal, obliging them frequently to break their line; and the positions116 they attacked had been previously fortified. The left could not see the right, or know what was passing there, for the rising ground between them. Every corps of the army fought apart, with unparalleled bravery, and ability too, but there was no cooperation in their efforts. The French were not then commanded by a General-in-chief, the resources of whose genius might have compensated for the advantages which the nature of the ground denied them and yielded to their enemies.

"The division of Lapisse first passed the ravine, attacked the fortified eminence, ascended it in defiance of a fire of grape-shot which mowed down its ranks, but was repulsed with the loss of its General and a great number of officers and soldiers. In retreating, it left the right of the fourth corps uncovered, which the British artillery took in flank, and forced for a moment to retire. The left of General Sebastiani's corps advanced under a most intense fire of artillery to the fort of a redoubt on the right of the English, and between the combined armies. It was too far advanced, and too soon forward—it was encountered and driven back by the united corps of the English right and the Spanish left. Assistance came, and the combat was renewed. In the center, Marshal Victor rallied the division of Lapisse at the foot of the hill and abandoned all further attempt to gain possession of it. The French then tried to turn it either to the right or left. Villatte's division advanced in the valley, and Ruffin's moved to the right of this by the foot of the Castilian mountains. The cavalry, forming a second line, were in readiness to debouch into the plain in the rear of the enemy whenever the infantry could open a passage.

"Just as the French began to move, the English, with two regiments of cavalry, made a charge against their masses. They engaged in the Valley, passed onwards regardless of the fire of several battalions of infantry, between the divisions of Villatte and Ruffin, and fell with impetuosity never surpassed on the 10th and 26th117 regiments of our chasseurs. The 10th could not resist the charge. They opened their ranks, but rallied immediately, and nearly the whole of the 23rd regiment of light dragoons at the head of the English cavalry was either destroyed or taken captive.

A division of the English Royal Guards,

Stationed on the left and centre of their army, being charged by the French, at first repulsed them vigorously. However one of its brigades, being too far advanced, was in its turn taken in flank by the fire of the French artillery and infantry, sustained a considerable loss, and retreated with difficulty behind their second line. The French took advantage of this success; they again moved forward, and but one other effort was necessary to break through into the plain, and combat on equal ground. But King Joseph thought it was too late to advance with the reserve, and the attack was delayed until the following day. 52 Night again closed over us, and the conflict ceased from exhaustion, without either side having won such a decided advantage as to entitle it to claim the victory.

"The corps of Marshals Victor and Sebastiani withdrew successively during the night towards the reserve, leaving an advanced guard of cavalry on the scene of the engagement, to take care of the wounded. The English, who expected a new attack in the morning, were greatly surprised when day dawned to see that their enemies, leaving twenty pieces of cannon, had retreated to their old position on the Alberche. The English and Spaniards, according to their own accounts, lost 6,616 men. 53 The French had nearly 10,000 slain."

Wellesley characterizes the battle as "a most dangerous one ... we had about two to one against us; fearful118 odds! but we maintained all our positions, and gave the enemy a terrible beating." Very few of the Spanish troops were engaged in any real sense, although those who took an active part behaved well, and one of the cavalry regiments "made an excellent and well-timed charge." The majority of them were in a "miserable state of discipline" and "entirely incapable of performing any maneuver, however simple." There was a sad lack of *morale*, qualified officers were few, and seemed either unable or unwilling to follow their allies in the matter of subjecting their men to specific regulations. When the British soldiers were engaged in removing the wounded and in burying the dead after Talavera, "the arms and accoutrements of both were collected and carried away by the Spanish troops."

The weak condition of his army prevented Wellesley from following the enemy, but as Venegas was on the move and threatening Madrid, this was not regarded as of supreme importance. Of more immediate concern was the alarming intelligence received by the Commander-in-Chief a few hours later that Soult's army was no longer in Galicia, but marching to intercept the British communications with Portugal.

CHAPTER XI
Wellesley's Defense of Portugal (1809–10)

"If I fail, God will, I hope, have mercy upon me, for nobody else will."

Wellesley.

Outline of specific point of views or inside knowledge in this chapter:

54. "Passages in the Early Military Life of General Sir George T. Napier, K.C.B." (London, 1884), pp. 111–12.
55. See *Chapterost XII.*
56. "Cambridge Modern History," vol. ix. p. 455. This authority gives the date of the battle of Tamames on the 18th October, but Wellington states that it occurred on the 19th.—See "Dispatches," vol. v. pp. 261 and 350.
57. Its object was to destroy the ships and dockyards at Antwerp.

Soult, joined by Mortier and Ney, had some 50,000 men with which to face the victor of Talavera. Had Cuesta guarded the mountain passes as he was supposed to do, Wellesley

Duke of Wellington

would not have found himself in so awkward a predicament. Both his front and rear were threatened, the former by Victor and Sebastiani and the latter by Soult. While his ranks were sadly depleted, those of the French were augmented. By great good fortune, General Craufurd and his great Light Division arrived on the morning of the 29th July, the day following the conclusion of the battle.

Rumor proved on this occasion a powerful ally of the Commander-in-Chief, for had not Craufurd heard of the supposed death of Sir Arthur, it is scarcely likely that he would have urged his men, each loaded with forty pounds' weight on his back, to march forty-three English miles in twenty-two hours. Wellesley made up his mind to advance against Soult but was forced to abandon the idea when he heard that the enemy had entered Plasencia in great force, thereby severing the British communications with Lisbon. A retreat "to take up the defensive line of the Tagus" became eminently necessary. "We were in a bad scrape," he writes on the 8th August, 120 "from which I think I have extricated both armies; and I really believe that, if I had not determined to retire at the moment I did, all retreat would have been cut off for both." "Both," of course, includes the Spaniards, whose "train of mismanagement," added to Soult's advance, were contributing causes of his withdrawal.

Duke of Wellington

The Spaniards promised to remain at Talavera to watch the movements of the enemy and to assist the wounded. No sooner was Wellesley out of the way than Cuesta felt that the position was untenable, with the result that many British soldiers, rendered unable to keep up with the Spanish troops because of their wounds, were made prisoners by Victor, who soon afterwards took possession of the town.

The Tagus was crossed at Arzobispo, and a rearguard of 8000 Spaniards, under Cuesta, left to defend the passage. At Almarez, the bridge of boats was broken to arrest Soult's advance from Naval Moral, no great distance away, and on the high road. As it happened, the French Marshal was able to cross the river at Arzobispo using a ford. He promptly defeated the Spanish force there and captured their guns. Not a few of the defenders fled, throwing away their arms and clothing, a conventional device. This was followed by the defeat of Venegas by Joseph and Sebastiani at Almonacid, near Toledo, where several thousand men were either killed, wounded, or captured, and of a Portuguese and Spanish column which had been detached from the main army, under Sir Robert Wilson, by Ney at the Puerto de Baños.

In the middle of August 1809

the various armies were occupying the following positions: British, Jaraicejo; Eguia, Deleytosa; Vanegas, La

Duke of Wellington

Carolina; Ney, Salamanca; Kellermann, Valladolid; Soult, Plasencia; Mortier, Oropesa and Arzobispo; Victor, Talavera and Toledo; Sebastiani, La Mancha.

The storm was followed by a lull, for the contesting armies were all but worn out and required rest. Wellesley[121] made his headquarters first at Deleytosa, and, when that place was vacated on the 11th, at Jaraicejo; the Spanish made the former town their headquarters, and the Portuguese army, under Beresford, withdrew within their home frontier.

"While the army remained in this position," namely, Deleytosa, General Sir George T. Napier records: "We suffered dreadfully from want of food. Nothing but a small portion of unground wheat and (when we could *catch them*) about a quarter of a pound of old goats' flesh each man; no salt, bread, or wine. As the Spaniards had plundered the baggage of the British army during the battle of Talavera, there was nothing of any kind to be procured to help us out, such as tea or sugar."[54] These defects the General determined to remedy, for "no troops can serve to any good purpose unless they are regularly fed," a maxim equivalent to that of Napoleon that "an army moves on its stomach."

Cuesta resigned his command,

Which was given to Eguia, but from henceforth the British Commander placed his sole reliance on his own forces. The lack of co-operation in the combined army was also evident in that of the French, for the several marshals had separated, and, to Napoleon's disgust, lost a golden opportunity of crushing the hated English, which was never again vouchsafed to them. "Force Wellesley to fight on every possible occasion," was his frequent cry. "Win if you can, but lose a battle rather than deliver none. We can afford to expend three men for every one he loses, and you will thus wear him out in the end." Wellesley preferred to conserve his energy, not to squander it.

After repeated requests for provisions and means of transport, all more or less evasively answered by the Spanish authorities, Wellesley carried out his threat and fell back upon the frontiers of Portugal. Not without a specific suggestion of irony, he was at this time appointed122 a Captain-General in the Spanish service, and received six Andalusian horses "in the name of King Ferdinand the VIIth." Shortly afterwards he was notified that he had been elevated to the Peerage, with the titles of Baron Douro of Wellesley, and Viscount Wellington of Talavera. From henceforth we shall style

Duke of Wellington

him Wellington, a signature he first adopted on the 16th September 1809.

"Nothing," he writes from Merida, on the 25th August, to Castlereagh, "can be worse than the officers of the Spanish army. It is extraordinary that when a nation has devoted itself to war. As this nation has, by the measures it has adopted in the last two years, so little progress has been made in any one branch of the military profession by any individual, and that the business of an army should be so little understood. They are really children in the art of war, and I cannot say that they do anything as it ought to be done, except running away and assembling again in a state of nature." In his opinion the Portuguese, under English officers, were better than the Spaniards, but both "want the habits and spirit of soldiers—the habits of command on one side, and of obedience on the other—mutual confidence between officers and men. Above all, a determination in the superiors to obey the spirit of the orders they receive, let what will be the consequence, and the spirit to tell the true cause if they do not. In short, the fact is, there is so much trick in the Portuguese army...."

At the beginning of September, he was at Badajoz, on the frontier, the advantage being, as Wellington says, "that the British army was centrally posted, about all the objects which the enemy might have in view. At any time, by a junction with a Spanish corps on its right, or a Portuguese or Spanish corps on its left, it could prevent the enemy from undertaking anything, excepting with a much larger force than they could allot to any one object." Here he heard that123 there was a likelihood of Soult attacking Ciudad Rodrigo. This information he obtained from an intercepted letter to Joseph. "The success of this scheme," he avers, "would do them more good, and the allies more mischief, than any other they could attempt; and it is most likely of all others to be successful." For this reason, he ordered Wilson to stay north of the Tagus so as to watch Soult's movements.

In the middle of the month the Spanish army of Estremadura, stationed at Deleytosa, was reduced to 6000 soldiers, the remainder, with Eguia, marching towards La Mancha. About the same time an army of some 13,000 men, under La Romana, whom Wellington describes as "more intelligent and reasonable" than most of his countrymen, moved from Galicia to the neighborhood of Ciudad Rodrigo. La Romana himself proceeded to Seville and was succeeded by the Duque del Parque, who marched towards Salamanca.

Wellington watched del Parque's forward movements with dismay, and ordered magazines to be prepared upon the

Douro and Mondego "to assist in providing for these vagabonds if they should retire into Portugal, which I hope they will do, as their only chance of salvation."

On the 10th, October Wellington was at Lisbon to arrange future operations, and where he studied "on the ground" the possibility of defending Portugal. This reconnaissance resulted in the defense known as the "Lines of Torres Vedras," of which particulars will be given as the story proceeds. 55

By the end of the month, he was back at Badajoz, writing endless dispatches about the thousand and one concerns—military, political and financial—of the two armies. By the beginning of November, del Parque was obliged to retire owing to the arrival from Estremadura of some 36,000 men under Mortier in Old Castile. Eguia's entry into La Mancha from Estremadura two months before had been followed by the 124 arrival of 30,000 of the enemy's troops under Victor in that province, after that the Spanish commander had withdrawn to the Sierra Morena, and the French to the Tagus.

The Spanish Government now entertained the hope of gaining the complete possession of Madrid. Two forces were to be honored with the carrying out of this ambitious project.

Duke of Wellington

The Army of La Mancha, now under the new General Areizaga, joined by the greater part of the Army of Estremadura, and consisting of 50,000 men, was to march from the Sierra Morena. Del Parque, with the army of Galicia, 20,000 strong, was to take Salamanca and then present himself before the capital. Areizaga met with some temporary success, but on the 19th November some 4000 of his men were either lying dead or wounded on the bloody field of Ocaña, within easy distance of Madrid. No fewer than 18,000 were taken prisoners by King Joseph's troops. When the distressed General gathered together the fragments of his shattered army in the Sierra Morena, only a half of the original number were present, which means that 3000 had deserted. He must have been sadly deficient in cannon, for the French had captured over fifty pieces.

Del Parque was scarcely more successful. He was attacked at Tamames on the 19th October by troops under Marchand, whom he defeated. Thus encouraged, he advanced to Salamanca, which the French had occupied, and taken possession of. In the last week of November, he was beaten at Alba de Tormes, to which he had retreated, with a loss of 3000 men. Some of his troops retired on Galicia, the remainder on Ciudad Rodrigo. 56

Duke of Wellington

With the object of giving the Spanish Government time to repair their losses in southern Spain, and surmising that whatever reinforcements the French might receive would be for use against the British now that the armies under Blake, Areizaga, and del Parque were scattered. Wellington prepared to move the greater part of his army north of the Tagus, towards the frontiers of Castile but leaving a body of troops under Lieutenant-General Hill at Abrantes, so that the Lower Tagus might not be left unguarded. Early in January 1810, Wellington made his headquarters at Coimbra, on the Mondego, and within a comparatively easy distance of the sea, conceivably a useful ally now that Napoleon was sending additional reinforcements, and Soult had 70,000 men at his disposal.

The passes of Despeña Perros, and Puerto del Rey through the Sierra Morena, but weakly defended by Spanish troops under Areizaga, were forced by the French without difficulty. On the last day of January 1810, Seville capitulated. That Wellington anticipated this is proved by a letter he wrote on that date to Lord Liverpool. Cadiz was saved from a similar fate by the Duke of Albuquerque, who reached the city in the nick of time, although Victor's outposts had been seen on the banks of the Guadalquivir. Major-General the Hon. W. Stewart was sent to assist in the defense of the place and arrived towards the end of February with some 5000 British and Portuguese troops while a British fleet lay in the Bay.

Meanwhile, Wellington was able to report considerable progress in some of the regiments in the Portuguese army, thanks very largely to the exertions of Marshal Beresford. Fifteen regiments he had seen while marching from Badajoz to Coimbra showed decided improvement in discipline, and he had "no doubt that the whole will prove a useful acquisition to the country." They were "in general unhealthy." The conduct of his own troops was "infamous" when not under the inspection of officers. "They have never brought up a convoy of money that they have not robbed the chest; nor of shoes, or any other article that could be of use to them, or could produce money, that they do not steal something."

126 The failure of the Walcheren Expedition 57 not only led to a duel between Canning and Castlereagh and the fall of Portland's administration but caused the British public to lose faith in things military. It seemed not at all improbable that the new Ministry formed by Perceval, in which Lord Wellesley became Foreign Secretary. Lord Liverpool became Secretary for War and the Colonies, would withdraw the British army from the Peninsula, especially as Talavera had aroused little or no enthusiasm, and a retreat was regarded by the man in the street as scarcely better than a defeat. In

Duke of Wellington

this connexion, it is interesting to note that when Wellington was asked what was the best test of a great general, he gave as his answer, "To know when to retreat; and to dare to do it." Aware of the state of public opinion, he did not press for further reinforcements.

In a letter to the Rt. Hon. John Villiers, dated Viseu, 14th January 1810, the Commander-in-Chief definitely states "that in its present state" the army was "not sufficient for the defense of Portugal." He anticipated having 30,000 active British troops when the soldiers then on their way from England and those in the hospital were available: "I will fight a good battle for the possession of Portugal, and see whether that country cannot be saved from the total wreck."

"I conceive," he concludes, "that the honor and interests of the country require that we should hold our ground here as long as possible. So please God, I will maintain it as long as I can. I will neither endeavor to shift from my own shoulders on those of the Ministers the responsibility for the failure, by calling for means which I know they cannot give, and which, perhaps, would not add materially to the facility of attaining our object. Nor will I give to the Ministers, who are not strong, and who must feel the delicacy of their own situation, an excuse for withdrawing the army from a position which, in my opinion, the honor and interest of the country require they should maintain as long as possible.

"I think that if the Portuguese do their duty, I shall have enough to maintain it. If they do not, nothing that Great Britain can afford can save the country. If I fail in saving it from that cause and am obliged to go, I shall be able to carry away the British army."

The war in Spain still continued see-saw fashion. A province would be apparently conquered by Napoleon's troops when no sooner did the troops march on than the trouble began again. This happened more especially with Suchet in Aragon. In Catalonia, the intrepid O'Donnell and his men flitted about like a will-o'-the-wisp and worked sad havoc whenever they came across a detached force though Hostalrich fell and Lerida surrendered in May, as well as the castle of Mequinenza a little later.

Napoleon

Had not been sleeping. In the previous year, he had been too occupied in humbling Austria and annexing Rome and the Ecclesiastical States to give much attention to the Peninsula. He now placed Marshal Masséna, Prince of Essling, who because of his success was known as "the spoilt child of victory"—incidentally he was the son of an inn-keeper—in command of 180,000 troops, for which purpose he arrived at

Valladolid in the middle of May 1810. Within a month, the French forces in Spain were raised to no fewer than 366,000 men of all ranks and arms.

Surely "the heir of Charlemagne," as Napoleon termed himself, was on the point of crushing the resistance of the Iberian Peninsula, and with it insignificant Portugal and Wellington? Was it feasible that 60,000 British, Germans, and Portuguese, sometimes aided but more often hindered by an "insurgent" rabble, and indirectly by the two remaining Spanish armies in Galicia and Estremadura, could contest with any likelihood of success more than a third of a million of trained troops? The law of probability answered in the negative.

CHAPTER XII
The Lines of Torres Vedras (1810)

"France is not an enemy whom I despise, nor does it deserve I should."

Wellington.

Outline of specific point of views or inside knowledge in this chapter:

58. General Sir George T. Napier, pp. 120–21.
59. Really his two reserve divisions, consisting of some 8000 men. See Oman, vol. iii. p. 132, and *post*, p. 139.
60. September 1810.
61. On the 27th September 1910, the centenary of the battle, an anniversary banquet was given at Busaco, which was attended by Wellington's grandson. King Manoel—now dethroned—signed a decree reaffirming the duke's Portuguese titles of Duke of Vittoria, Marquis of Torres Vedras, and Count of Vimiero. Celebrations were also held on the site of the battle.

Pasquier, who had the privilege of knowing most of the generals of the Revolution and of the Empire, says of Masséna that he was "France's first military commander after Napoleon." Neither Pichegru, Moreau, Kléber, nor Lannes gave the Chancellor "as completely as Masséna, the idea of a born warrior, possessing a genius for war, and endowed with all the qualities which render victory certain. His eagle eye seemed made to scan a field of battle. One could understand, on seeing him, that the soldier under his command never believed it was possible to retreat."

Figure 15 Renault - André Masséna, duc de Rivoli, prince d'Essling, maréchal de France (1756-1817)

Duke of Wellington

Masséna's first imperative operation in the Peninsula was the siege of the Spanish fortress of Ciudad Rodrigo. Although Wellington was in the neighbourhood, he was not to be enticed away from his immediate objects, which were the defense of Lisbon and the complete organization of the army for service when the action became absolutely imperative. Notwithstanding an outstanding defence for over two months on the part of Governor Herrasti, the garrison of Ciudad Rodrigo was compelled to surrender on the 10th July. In August, Masséna crossed the frontier preparatory to beginning the siege of129 Almeida, near the river Coa, next to Elvas the strongest place in Portugal.

On the 24th July, Craufurd and his famous Light Division—not Light Brigade as some would have it—had a fierce tussle with Ney's corps of 24,000 men. Craufurd, who had only 4000 troops at his disposal, entertained no wild notion of preventing the investment of the place, but as he was suddenly attacked he was obliged to fight. Had he been a more cautious soldier he would have crossed the Coa before Ney came up, as Wellington had suggested on the 22nd. Indeed, so early as the 11th, the Commander-in-Chief had said, "I would not wish you to fall back beyond that place (*i.e.* Almeida) unless it should be necessary. But it does not appear necessary that you should be so far, and it will be safer that you should be nearer, at least with your infantry." He delayed too late, and

Duke of Wellington

thereby lost over 300 men. While the last of the soldiers were crossing the bridge which spanned the swollen river, for it had rained in torrents the previous night, a lanky Irish lad of nineteen years, named Stewart, and known by the 43rd as "The Boy," positively refused to pass over. "So this is the end of our boasting! This is our first battle, and we retreat! The Boy Stewart will not live to hear that said," he cried, and turning back he slashed at the oncoming French until he fell dead. Even more courageous was the conduct of Sergeant Robert M'Quade, five years Stewart's senior. He happened to catch sight of two French soldiers with levelled muskets awaiting the British to ascend a bank. A boy of sixteen, afterwards famous as Sir George Brown (Colonel-in-Chief of the Rifle Brigade) was on the verge of being shot by them when the sergeant pulled him back from the fatal spot. "You are too young, sir, to be killed," he cried as his own body received two bullets and fell in a lifeless heap at the feet of the youth.

That Colonel Cox, who was in charge of the fortress, would have stayed Masséna's advance for a considerable time is extremely likely, but, unfortunately, he was not given the opportunity to display his prowess. The powder magazine blew up, almost destroying the town and necessitating immediate surrender. The pursuit of Wellington, "to drive him into the sea," seemed a comparatively easy task until the advance showed that the British General had caused the country to be stripped almost entirely of provisions. Thus Napoleon's policy of making "war support war" by plundering and raiding the enemy's country, completely broke down. "In war all that is useful is legitimate," he says, and Wellington had followed the maxim, after having obtained permission for the destruction of provisions from the Portuguese Regency, which included Mr. Charles Stuart, the British Minister at Lisbon. What Wellington's measures meant to Masséna's army is summed up in a single sentence by Sir Harry Smith, who carried a dispatch to Lord Hill through territory occupied by the enemy. "The spectacle," he says, "of hundreds of miserable wretches of French soldiers on the road in a state of *starvation* is not to be described." Nor was this all. Not only did the place resemble a desert in the difficulty of obtaining means of sustenance, but the majority of the inhabitants had fled, some seeking the fastness's of the mountains, others the larger cities such as Lisbon and Oporto.

As Masséna advanced so, Wellington retreated towards the important lines of Torres Vedras, upon the construction of which thousands of peasants, under the direction of British engineers, had been busy for six months.

These impressive defenses are thus described by one who knew them. 58 They "consisted of redoubts and field-works of various kinds; according to the ground they were to defend, and all connected with each other by entrenchments, etc., so that, when occupied by the army, it would almost be impossible to force them. But, even supposing this first line of defense should be131 carried by the enemy, there was another, much more contracted, to retreat upon, where an outside force could hold out against the French army and cover the embarkation of the British, should Lord Wellington be at last required to quit Portugal. I cannot help considering this retreat to the lines, and the pertinacity with which he held them in spite of every difficulty. The remonstrances of the Government at home, which was seized with alarm, as the greatest proof of a mastermind and genius that could be given. That proved Lord Wellington to be superior to any general the French had, except Napoleon. In short, that he was, next to Buonaparte himself, the first general of the day. And I am further convinced that, had he the same opportunities that Napoleon had, he would have proved as great a general, as his capacity and powers of mind would have strengthened and expanded in proportion to the vastness of his views and the obstacles to be surmounted."

An officer of the 60th Rifles, who served behind them, furnishes a more detailed pen-sketch. "The line of defense was double," he writes. "The first, which was twenty-nine miles

Duke of Wellington

long, began at Alhandra, on the Tagus, crossed the valley of Armia. Which was rather a weak point, and passed along the skirts of Mount Agraça, where there was a large and strong redoubt. It then passed across the valley of Zibreira, and skirted the ravine of Runa to the heights of Torres Vedras, which were well fortified; and from thence followed the course of the little river Zizandre to its mouth on the sea-coast. The line followed the path of the mountain track which extends from the Tagus to the sea, about thirty miles north of Lisbon. Lord Wellington's headquarters were fixed at Pero Negro, a little in the rear of the center of the line, where a telegraph was secure corresponding with every part of the position. The second line, at a distance varying from six to ten miles in the rear of the first, extended from Quintella, on the Tagus, by Bucellas, Montechique, and Mafra, to the mouth of the little river S. Lourenço, on the sea-coast, and was twenty-four miles long. This was the stronger line of the two, both by Nature and Art, and if the first line were forced by the enemy, the retreat of the army upon the second was secure at all times. Both lines were secured by breastworks, abattis, stone walls with banquettes, and scraps. In the rear of the second line, there was a line of embarkation, should that measure become necessary, enclosing an entrenched camp and the fort of St Julian." As many as 120 redoubts and 427 pieces of artillery were scattered along these lines. "Lord Wellington had received reinforcements from England and Cadiz. The Portuguese army had also been strengthened, and the Spanish division of La Romana, 5000 resilient, came from Estremadura to join the Allies. **59** So that the British commander had about 60,000 regular troops posted along the first and second lines, besides the Portuguese militia and artillery (which manned the forts and redoubts and garrisoned Lisbon). This was a satisfactory body of English marines which occupied a line of embarkation, a powerful fleet in the Tagus, and a flotilla of gunboats flanking the right of the British line. It was altogether a colossal line of defense, conceived by the military genius of the British commander, and executed by the military skill of the British engineer officers."

Wellington continued to fall back until he reached "Busaco's iron ridge," north of the Mondego. Here he determined to offer Masséna battle, for three principal reasons. First, there was a growing discontent amongst the rank and file of his army because of lack of active warfare and the fall of Ciudad Rodrigo and Almeida, and a victory would put an end to this growing despondency. Second, also a military consideration, the orders he had given for the laying waste of the districts about Lisbon were not yet fully carried out. Third, from a political point of view it was necessary because 133 it would show that he was not about to lock himself up within the lines of Torres Vedras because he was incapable or afraid of Napoleon's legions. In a word, it would "restore confidence," a matter of first importance. It is somewhat incorrect to term Busaco a "useless battle" as some historians have done.

"On the 25th and 26th," [60] says M. de Rocca, "the French corps arrived successively at the foot of the mountains Sierra de Busaco, whose summits they found occupied by the Anglo-Portuguese army. At six o'clock, on the morning of the 27th, they marched in the column against the right and center of that army, on the two roads leading to Coimbra, by the village of San Antonio de Cantaro, and by the convent of Busaco. These roads were cut up in several places and defended by artillery. The mountain over which they pass is beside encumbered with steep rocks and is tough to access.

"The French column which attacked the right of the English advanced with intrepidity, in spite of the fire of their artillery and light troops. It reached the top of the eminence after sustaining a considerable loss, and began to deploy in line with the greatest coolness, and most perfect regularity. But a superior force again assaulted it and compelled it to retire. It soon rallied, made a second attack, and was again repulsed. The French battalions, which advanced against the convent of Busaco, where the left and center of the English divisions joined, were also driven back, a little before they reached that post. General Simon, who had been struck by two balls during the charge, was left on the height, and a great many wounded officers and soldiers.

"The position occupied by the English and Portuguese on the brow of the hill, formed the arc of a circle, whose two extremes embraced the ground over which the French had to advance. The allied army saw the least movements made below them and had time to form to receive any powerful body before it arrived. This circumstance134 materially contributed to the advantage they obtained....

"Marshal Masséna judged that the position of Lord Wellington could not be carried in front, and resolved to turn it. He kept up an indiscriminate fire till the evening and sent off a body of troops by the mountain-road, which leads from Mortago to Oporto. The English and Portuguese, in consequence of this movement, abandoned their position on the mountain of Busaco."

The attack on the British left was led by Ney, and it succeeded in driving in the sharp-shooters. The French had practically reached the summit, as Rocca states, when Craufurd's division, concealed in a hollow, gave them the full benefit of their fire. "The enemy," says Sir Charles Stewart, who fought on this memorable day, "unable to retreat, and afraid to resist, were rolled down the steep like a torrent of hailstones driven before a powerful wind. Not the bayonets only, but the very hands of some of our brave fellows, became in an instant red with the blood of the fugitives. More brilliant or more certain charges than those executed this day by the two divisions which bore the brunt of the action were never perhaps witnessed; nor could anything equal the gallantry and intrepidity of our men throughout, except imaginable the hardihood which had ventured upon so desperate an attack."

Reynier's two divisions, 15,000 men in all, attacked Picton's 3rd division on the right. The troops of Generals Hill and Leith, moving rapidly to Picton's aid, decided their fate. "The right of the 3rd division had been, in the first instance, borne back," says an eye-witness, "the 8th Portuguese had suffered most severely. The enemy had formed, in good order, upon the ground which they had so boldly won, and were preparing to bear down to the right, and sweep our field of battle. Lord Wellington arrived on the spot at this moment and aided the gallant efforts of Picton's regiments, the fire of whose musketry was terrible, by causing two guns to play upon the French flank with grape. Unshaken even with this destruction, they still held their ground, till, with levelled bayonets and the shout of the charge, the 45th and 88th regiments, British, most gallantly supported by the 8th

Portuguese, rushed forwards, and hurried them down the mountainside with a fearful slaughter."

"This movement," writes Wellington, "has afforded me a favorable opportunity of showing the enemy the description of troops of which this army is composed. It has brought the Portuguese levies into action with the enemy for the first time in an advantageous situation. They have proved that the trouble which has been taken with them has not been thrown away and that they are worthy of contending in the same ranks with British troops in this fascinating cause, which they afford the best hopes of saving.

"Throughout the contest on the Serra, and in all the previous marches, and those which we have since made, the whole army have conducted themselves in the most regular manner. Accordingly, all the operations have been carried on with ease; the soldiers have suffered no privations, have undergone no unnecessary fatigue, there has been no loss of stores, and the army is in the highest spirits."

The total British and Portuguese losses, according to the official figures, were 197 killed, 1014 wounded, and 58 missing. Masséna reported casualties to the number of 4486 men, including five generals. Anything but a pleasant feeling existed between the French Commander-in-Chief and Ney previous to the battle; the result merely deepened their unfriendliness, a low contrast to the cordial relations between Wellington and his colleagues. 61

136 It is both delightful and pathetic to know that, after the last roll of the guns had echoed through the valley, the British and the French put aside their weapons and worked side by side in the humanitarian task of searching for the wounded. It was the final scene of the tragedy, acted after the curtain had fallen. It is recorded, as one of the incidents, that a German officer serving with Napoleon's colors, who had a brother in the British 60th Regiment, asked a sworn enemy of an hour ago if he knew what had happened to his relative? He answered his own pathetic question by finding the soldier's corpse.

Books may tell of its story, but only the heart can know How war is robbed of its glory, By the brave ones lying low,

Duke of Wellington

CHAPTER XIII
Masséna beats a Retreat (1810–11)

"There will be a breeze near Lisbon, but I hope we shall have the best of it."

Wellington.

Outline of specific point of views or inside knowledge in this chapter:

62. The writer is speaking literally.
63. The usual French mode of attack.
64. Not Marshal Soult, but his nephew.
65. The Proclamation is printed in full in Gurwood's edition of "Wellington's Dispatches," vol. vii. pp. 455–7.
66. Lady Butler's picture, "Steady, the Drums and Fifes," represents this regiment drawn up on the ridge.

Owing to the failure of one of Wellington's officers to occupy the Boialva Pass, Masséna was able to turn the British position, with the result that his advanced guard appeared in front of Coimbra on the evening of the 30th September.

When the Commander-in-Chief saw the French army defiling across the mountains "he seemed uneasy," according to one who watched him, "his countenance bore a fierce, angry expression, and, suddenly mounting his horse, he rode away without speaking."

No attempt was made to attack the enemy, Wellington considering it more prudent to leave the ridge, cross the Mondego, and retreat towards Lisbon. This resolution came to on the 28th September, and on the 1st October the last man of the rear-guard had evacuated the town. "Although I could not save Coimbra," Wellington writes, "I have little doubt of being able to hold this country against the force which has now attacked it."

The place was immediately occupied by Masséna's famished troops, who found it not entirely destitute of eatables, as seemed only too probable judging by previous138 experience, although much of the food had been destroyed by Wellington's orders. They had merely to help themselves to what they could find, for most of the population had followed in the wake of the allied army. "The inhabitants of the country have fled from their houses universally," the Commander-in-Chief writes to the Earl of Liverpool from Alcobaço on the 5th October, "carrying with them everything they could take away which could be deemed useful to the enemy. The habits of plunder which have been so long encouraged in the enemy's army prevent them from deriving any general advantage from the little resource which the inhabitants may have been obliged to leave behind them."

It is the straightforward utterance of a sincere man. Wellington seldom indulged in colorful language; he had neither the natural ability which commands a tactical choice of language nor the time for clear diction. His mind was cast in a sterner mold; he craved for exactness, for cold, rational calculations, for ungarnished essentials.

For graphic details, we must turn to such an authority as Sir Charles Stewart, who writes with the fluency of a gifted war-correspondent permitted to ride with the officers and obtain a view of everything of importance. "Crowds of men, women, and children," he says, "—of the sick, the aged, and the infirm, as well as of the robust and the young—covered the roads and the fields in every direction. Mothers might be seen with infants at their breasts hurrying towards the capital, and weeping as they went. Old men, scarcely able to totter along, made way chiefly by the aid of their sons and daughters; while the whole wayside soon became strewed with bedding, blankets, and other species of household furniture, which the weary fugitives were unable to carry farther. During the retreat of Sir John Moore's army, numerous heartrending scenes were brought before us. For then, as now, the people, particularly139 in Galicia, fled at our approach; but they all returned sooner or later to their homes, nor ever dreamed of accumulating upon our line of march, or following our fortunes. The case was different here. Those who forsook their dwellings deserted them under the persuasion that they should never behold them again; and the agony which such an apprehension appeared to excite among the majority exceeds any attempt at description.... It could not but occur to us that, though the devastating system must inevitably bear hard upon the French. The most serious evils would, in all probability, arise out of it, both to ourselves and our allies, from the famine and general distress which it threatened to bring a crowd so dense, shut up within the walls of a single city. At the moment, there were some standing amongst us who seemed not disposed to view it with reprobation; because, while they condemned its apparent violation of every feeling of humanity and justice, they doubted the soundness of the policy in which it originated."

Leaving a meagre force to guard the wounded and sick at Coimbra, Masséna started off in pursuit of the enemy as soon as the most primeval of creature comforts had been satisfied. Six days after his soldiers had left the place, namely, the 11th October 1810, Wellington's men entered the lines of Torres Vedras, but so rapid had been the French advance that they began to appear on the following morning. La Romana had crossed from Estremadura with several thousand Spanish troops, thereby adding to Wellington's forces, while Portuguese militia threatened the enemy's communications.

Masséna dared not attempt to retreat for fear of incurring Napoleon's displeasure. His only hope, as he repented at leisure, was that the supplies of the defenders might fail, or that the Emperor, in response to urgent dispatches, would speedily send reinforcements both of men and of arms. The news that Coimbra, its garrison, and its invalids, had fallen into the hands of militia under Colonel Trant merely added insult to injury. As regards140 "starving out" the British and their allies, it was far more probable that their own food would run out, for while Wellington held the Tagus a constant supply of the necessaries of life was secured from incoming ships. Hunger did indeed eventually drive Masséna from Santarem, a town some thirty miles from Wellington's lines, on which he had been forced to fall back in November. The place, perched on the summit of a height between the rivers Rio Mayor and Aviella, was admirably suited for defensive purposes, but after the surrounding country had been stripped there was nothing to do but retire. The Marshal was fortunate in finding a district which the Portuguese had not laid bare.

It sounds almost incredible, but it is recorded that when Masséna succeeded in crossing the frontier his men were so famished that one of them consumed no less than seventeen pounds of homemade bread. The French General awaited with feverish anxiety the coming of Soult to his relief for nearly four months. That worthy was fighting battles and besieging Badajoz, which the Spaniards surrendered on the 10th March 1811, five days after his colleague had been forced by sheer necessity to begin a retreat across the mountains towards Ciudad Rodrigo.

The author of "The Journal of an Officer," a reliable eye-witness, thus describes the town after Masséna had left it: "I have been for some weeks given Santarem, and saw at last with pleasure some symptoms of the French abandoning it. The first was setting fire to one of the principal convents in the better part of town, and part of the lower town; the volume of smoke was immense for three days. On the fourth morning, some information to depend on reached us, and the bugle of attack roused us from our pillows. The haze of the morning clearing up, we could easily perceive the out sentinels were men of straw, and quite passive. 62 In fact, a better-managed retreat was never executed.141 Not a vestige of a dollar's worth remained. Being at the outposts with the 11th Dragoons and the 1st Royals, I entered with them, and three miserable deserters, who had hidden with one too ill to move, were the only enemies to be found. Such a scene of horror, misery, and desolation, scarce ever saluted the eye of man. Smoking ruins, the accumulated filth of months, horses and human beings putrefied to suffocation, nearly caused to many a vomiting. The houses had hardly a trace of wood, doors, windows, ceilings, roofs, burnt. Where the sick had expired, there left to decay! The number thus left were great. Every church demolished, the tombs opened for searching after hidden plate, every altar-piece universally destroyed, and the effluvia so offensive as to defy describing.

"In some gardens, the miserable heads, undecayed, stuck up like scarecrows; in some wells, a body floating.

"Down a precipice to which we were invited by the prospect to look, the human and animal carcasses ... repulsed our senses, and shudderingly vibrated the soul at the savage, horrible, evil acts of a French army. Greater spirits, better discipline, and more order, never attended an army than this. But to see the country, is to weep for the horrors of war. Such gross excess I never saw before. Every town, village, or cottage is destroyed. The growing nursery and the wild grove, each havocked for destruction's sake. The pot that refined the oil broken, the wine-press burnt, for burning's sake. The grape vines destroyed as noxious weeds. The furniture unburnt thrown from the windows, and with carriages, etc., made a bonfire of. The large libraries strewed over the land in remnants of paper. The noble convent in ashes, and the underprivileged, unhappy, aged inhabitants, unable to flee, hung around as ornamenting the walls, ten or twelve in a place!"

Wellington, who had now received reinforcements, moved his headquarters to Santarem on the 6th March, 142 anxious to overtake the enemy with the least possible delay. He received the usual conflicting accounts of the direction taken by them and their possible destination. Oporto was suggested, which the Commander did not believe, "but they are in such a state of distress, that it may be expected that they will try anything, however desperate. But I follow them closely; and they will find it difficult to stop anywhere, for any purpose, till they shall draw near the frontier." He detached two divisions under Beresford, hoping that he might be able to relieve Badajoz, and with five others continued to keep "close at their heels," to use his own expression. Unfortunately, the place fell before it was possible for Beresford to reach it. Had the Governor held out, Wellington was of the opinion that "the Peninsula would have been safe," and the relief of the south of Spain practically assured.

Duke of Wellington

"Affairs" with the enemy were frequent during Wellington's pursuit, but by forcing them to evacuate the various positions they attempted to occupy, such as Pombal, Redinha, Cazal Nova and Foz d'Aronce. Any designs they might have had against the northern provinces were prevented, notwithstanding the fact that the country afforded "many advantageous positions to a retreating army, of which the enemy have shown that they know how to avail themselves."

In writing to the Earl of Liverpool, Wellington remarks that "their conduct throughout this retreat has been marked by a barbarity seldom equaled, and never surpassed." He tells a moving story of plunder, the burning of houses, a convent, and a bishop's palace. "This is the mode," he adds in a burst of indignation, "in which the promises have been performed, and the assurances have been fulfilled, which were held out in the proclamation of the French Commander-in-Chief, in which he told the inhabitants of Portugal that he was not come to make war upon them, but with a powerful army of 110,000 men to drive the English into the sea.

"It is to be hoped that the example of what has occurred in this country will teach the people of this and of other nations what value they ought to place on such promises and assurances. That there is no security for life, or for anything which makes life valuable, excepting in decided resistance to the enemy."

The difficulties of the chase were many and often almost insurmountable. Boats and bridge-building materials were scarce and caused a delay in crossing rivers. Shoes wore out rapidly on account of the bad quality of the leather, and many of them were too small. Endless trouble was caused by the Spanish muleteers, who absolutely refused to attend the Portuguese troops, some of whom Wellington was obliged to leave in the rear owing to the scarcity of provisions. For instance, two brigades of infantry had to make nine days' provisions, consisting chiefly of bread and a little meat supplied by the British commissariat, last for twenty-four days. "This is the assistance I receive from the Portuguese Government!" the Commander-in-Chief writes, and one can imagine his grim face hardening as he pens the words. There were the usual grievances against the rascally army contractors. The boots sent out were of bad quality, "in general too small." We find him ordering 150,000 pairs of boots and 100,000 pairs of soles and heels at a time.

The most serious action during Masséna's retreat was fought at Sabugal, on the Coa, on the 3rd April. "We moved on the 2nd," Wellington says when giving details of the engagement to Beresford, "and the British army was formed opposite to them; the divisions of militia, under Trant and Wilson, were sent across the river at Cinco Villas, to alarm Almeida for its communication. Yesterday morning"—he is writing on the 4th inst.— "We moved the whole army (except the 6th division, which remained at Rapoula de Coa, opposite Loison) to the right, to turn this position and force the passage of the river. The 2nd corps could not have stood here for a moment. Unfortunately the Light division, which formed the right of the whole, necessarily passed first, and the leading brigade, Beckwith's, drove in the enemy's piquets, which were followed briskly by four companies of the 95th, and three of Elder's caçadores, and supported by the 43rd regiment. At this time, there came on a rain storm, and it was as difficult to see as in the fogs on Busaco, and these troops pushed on too far and became engaged with the main body of the enemy. The light infantry fell back upon their support, which instead of halting, moved forward. The French then seeing how weak the body was which had passed, attempted to drive them down to the Coa, and did oblige the 43rd to turn. They rallied again, however, and beat in the French; but were attacked by fresh troops and cavalry, and were grateful to retire; but formed again, and beat back the enemy. At this time, the 52nd joined the 43rd, and both moved on upon the enemy, and to be charged and attacked again in the same manner, and beat back. They formed again, moved forward upon the enemy, and established themselves on the top of the hill in an enclosure, and here they beat off the enemy.

"But Reynier was placing a body of infantry on their left flank, which must have destroyed them, only that at that moment the head of the 3rd division, which had passed the Coa on the left of the Light Division, came up, and opened their fire upon this column. The 5th division, which passed this bridge and through this town [Sabugal], made their appearance.

"The enemy then retired, having lost in this affair a howitzer, and I should think not less than 1000 men.

"Our loss is much less than one would have supposed possible, scarcely 200 men. The 43rd have 73 killed and wounded. But really these attacks in columns against our lines are very contemptible.

145 "The contest was latterly entirely for the howitzer, which was taken and retaken twice, and at last remained in our hands. Our cavalry, which ought to have crossed the Coa on the right of the Light Division, crossed at the same ford, and, therefore, could be of no use to them. Besides they went too far to the right.

Duke of Wellington

"In short, these combinations for engagements do not answer, unless one is on the spot to direct every slight movement. I was upon a hill on the left of the Coa, immediately above the town, till the 3rd and 5th divisions crossed, whence I could see every movement on both sides, and could communicate with ease with everybody; but that was not near enough.

"We took 6 Officers, and between 200 and 300 prisoners, and Soult's 64 and Loison's baggage."

Two days after this affair on the banks of the Coa Masséna crossed the frontier, having been literally driven out of Portugal. Within a few hours, we find Wellington urging on Beresford the necessity for a strict blockade of Badajoz preparatory to besieging it. Masséna fell back upon Salamanca while Wellington busied himself with the investment of Almeida, where a French garrison had been left. With Ciudad Rodrigo, the second and remaining place occupied by the Marshal's troops, he felt he could do little at the moment beyond intercepting supplies. These two forts, which are within comparatively easy distance and almost parallel, the one in Portugal and the other in Spain, were paramount, and commanded the north-eastern frontier of the former country.

Duke of Wellington

Incidentally, the British Commander-in-Chief also took the opportunity to publish a lengthy Proclamation to the Portuguese nation, of which the following is a brief synopsis. He informs the inhabitants that they are now "at liberty to return to their occupations." That nearly four years have elapsed since "the tyrant of Europe" invaded the country, the object being "the insatiable desire of plunder, the wish to disturb the tranquillity, and to enjoy the riches of a people who had passed nearly half a century in peace." He then strikes a deeper note and adds a few words of advice as to the future:

The Marshal General,

However, considers it his duty, in announcing the intelligence of the result of the last invasion, to warn the people of Portugal, that, although the danger is removed, it is not entirely gone by. They have something to lose, and the tyrant will endeavor to plunder them: they are happy under the trivial government of a beneficent Sovereign, and he will endeavor to destroy their happiness: they have successfully resisted him, and he will struggle to force them to submit to his iron yoke. They should be unremitting in their preparations for decided and steady resistance. Those capable of bearing arms should learn the use of them. That whose age or sex renders them unfit to bear arms should fix upon places of security and concealment, and should make all the arrangements for their easy removal to them when the moment of danger shall approach. Valuable property, which tempts the avarice of the tyrant and his followers, and is the great object of their invasion, should be carefully buried beforehand, each individual concealing his own, and thus not trusting to the weakness of others to keep a secret in which they may not be interested.

"Measures should be taken to conceal or destroy provisions which cannot be removed, and everything which can tend to facilitate the enemy's progress; for this may be depended upon, that the enemy's troops seize upon everything and leave nothing for the owner.

"By these measures, whatever may be the superiority of numbers with which the desire of plunder and of revenge may induce, and his power may enable, the tyrant again to invade this country, the result will be firm. The independence of Portugal, and the happiness of its inhabitants will be finally established to their eternal honor." 65

However, "beneficent" the Sovereign—who was a lunatic and out of the country—might be, Wellington had little that was good to say of its present rulers. He told them that he would inform the home Cabinet "that they cannot with propriety continue to risk a British army in this country unsupported by any exertion of any description on the part of the Portuguese Government." The army was lamentably deficient "in that essential arm, its cavalry," and the commissariat arrangements remained hopelessly under provided.

The blockade of Almeida being "a simple operation, which I do not think the enemy have the means or inclination to interrupt," Wellington left it in the hands of Lieut.-General Sir Brent Spencer in the middle of April, and set out from Villa Fermosa for Alemtejo to discuss his future projects with Castaños and also to visit Beresford. He knew that the French at Almeida would be forced to withdraw or surrender owing to the scarcity of provisions, but at Ciudad Rodrigo "there is a respectable garrison, and we certainly shall not get that place without a siege; for which God knows if we shall have time before the enemy will be reinforced. The first object is certainly Badajoz, and, as soon as I recognize whether any or what part of our train is required for the attack on that place, I shall send the remainder to Oporto, and make all the arrangements for the eventual attack of Ciudad Rodrigo."

Duke of Wellington

As Soult was then busily occupied in fortifying Seville, to the south of Badajoz, the siege of the latter city became imperative, and without unnecessary delay. Soult might attempt to relieve Badajoz; without a doubt his presence at Seville precluded the likelihood of the garrison being deceived by any feint or actual attack made on that place by the allies with the object of distracting their attention.

148 Although Wellington did not meet Castaños personally during his visit to the South, he sent him a plan of operations, to be undertaken with Blake and Ballasteros in co-operation with Beresford, and got through an immense amount of work in connection with the siege. "The continued and increasing inefficiency of the Portuguese regiments with this army," gave him much cause for concern. On the 30th April 1811, four days after Parliament had thanked him for the liberation of Portugal, he tells Beresford that "if some strong steps are not taken, the Portuguese force with this part of the army (*i.e.* Wellington's) will be annihilated." He concludes by saying that he must report the matter to the home authorities, which he did. "The Ministers and the English public believe that we have 30,000 men for whom we pay, and half as many more supported by the Portuguese Government. I do not trust that I have here 11,000, or that you have 5000, and of the number many are not fit for service."

Masséna was not the type of man who without difficulty acknowledges defeat. He had been busily engaged at Salamanca in getting what remained of his army into working order. He had lost at least 25,000 of the 70,000 men who had entered Portugal, but when he decided to go to the assistance of Almeida he could with difficulty muster only 39,000, some 5000 more than Wellington could put into the field. Having relieved Ciudad Rodrigo, Masséna crossed the Agueda, with the fixed intention of raising the blockade of Almeida. On the 3rd May he was in sight of the British army, now arrayed at Fuentes de Oñoro.

The Commander-in-Chief had returned from his travels on the 28th of the previous month, after having been informed by Spencer of the gathering of the enemy. "I'll venture to say," remarks Kincaid, "that there was not a heart in the army that did not beat more lightly when we heard the joyful news of his arrival the day before the enemy's advance." On the 3rd May, the British[149] were "warmly but partially engaged," and "made no progress in raising the blockade."

The real battle began on the 5th, and was, in Alison's opinion, "the most critical in which Lord Wellington was engaged in the whole war, and in which the chances of irreparable defeat were most against the British army." He then gives some of Sir Charles Stewart's reflections on the fight, which help us to appreciate its difficulties from an actual eye-witness who took a leading part in the battle. "Masséna's superiority to us," he notes, "both in cavalry and artillery, was very great; while the thick woods in our front afforded the most suitable plateau which he could have desired for the distribution of his columns unseen, and therefore disregarded. Had he rightly availed himself of this advantage, he might have poured the mass of his force upon any single point, and perhaps made an impression before we could have had time to support it. Had he commenced his attack with a violent cannonade, it must have produced some havoc, and probably considerable confusion, in our line. He might then have moved forward his cavalry *en masse*, supporting it by resilient columns of infantry; and had either the one or the other succeeded in piercing through, our situation would have been by no means a lucky one.... Had he thrown his cavalry round our right flank—a movement which we should have found it no easy matter to prevent—crossed the Coa, advanced upon our lines of communication. He could have stopped our supplies, at the moment when, with his infantry, he threatened to turn us. He then pushed upon Sabugal and the places near, he might have compelled us to pass the Coa with all our artillery at the most disadvantageous places, and cut us off from our best and safest retreat. There was, indeed, a time during the affair of the 5th, when his design of acting in this manner was totally apprehended; and Lord Wellington was in consequence reduced to the necessity of deciding whether he should relinquish the Sabugal road or raise the150 blockade of Almeida. But Lord Wellington's presence of mind never for a moment forsook him. He felt no distrust in his troops. He retains his hold on a secure and accessible line of retreat was therefore to him a consideration of less moment than to continue an operation of which the ultimate success could now be neither doubtful nor remote. He at once determined to expose Sabugal rather than throw open a communication with Almeida. It was a bold measure, but it was not adopted without due consideration, and it received an ample reward in the successful termination of this hard-fought battle."

Wellington's line was extended on a table-land between the rivers Turones and Dos Casas. It reached several miles, namely, from Fort Conception, which covered Almeida (opposite the village of that name he disposed of his center), to beyond Nava d'Aver, his rational being at Fuentes de Oñoro. Poço Velho, between the last place and Nava d'Aver, was also occupied by the left wing of the 7th Division, commanded by General Houstoun.

Masséna's first movement was to attack the Spanish irregulars, under Don Julian Sanchez, stationed on the hill of

Duke of Wellington

Nava d'Aver, which was neither a lengthy nor a tricky process.

Major-General Houstoun scarcely fared better, two of his battalions being routed. The immediate consequence was that Captain Norman Ramsay's battery of Horse Artillery, which were supporting Houstoun, were soon fighting against fearful odds. Using a magnificent charge, while the attention of part of the French force was detracted by the dragoons under Sir Stapleton Cotton, Ramsay made good his escape with every gun.

The situation was extremely critical when the squares of the 7th and Light Divisions were attacked by the enemy's cavalry, but Wellington did not hesitate for a moment as to the best course to pursue. He abandoned Nava d'Aver and closed in his line by a complete change of front, withdrawing some of his divisions to the heights,151 and Houstoun's men behind the Turones, to a position near Freneda, which became the British right and Fuentes de Oñoro the left.

"Montbrun's cavalry," we are told, "merely hovered about Craufurd's squares. The plain was soon cleared, the cavalry took post behind the centre, and the Light Division formed a reserve to the right of the 1st Division, sending the riflemen among the rocks to connect it with the 7th Division, which had arrived at Freneda, and was there joined by Julian

Duke of Wellington

Sanchez. At the sight of this new front, so deeply lined with troops, the French stopped short, and commenced a heavy cannonade. Which did great execution, from the closeness of the allied masses. Twelve British guns replied with vigour, and the violence of the enemy's fire abated; their cavalry then drew out of range, and a body of French infantry, attempting to glide down the ravine of the Turones, was repulsed by the riflemen and light companies of the Guards."

Meanwhile, a fierce conflict was taking place in the village of Fuentes. It continued see-saw fashion until the evening, both sides bringing up reserves and contesting every inch of the ground. Three regiments were driven from the lower parts of the village, but reinforcements were at hand, and the higher streets were never abandoned, although a chapel held by the troops in that quarter was evacuated. At nightfall, the French crossed the river, leaving 400 of their dead in the village. Wellington averred that the battle "was the most challenging I was ever concerned in, and against the greatest odds. We had very nearly three to one against us engaged; about four to one of cavalry; and, moreover, our cavalry had not a gallop in them while some of that of the enemy was fresh, and in excellent order. If Bony had been there, we should have been beaten."

Duke of Wellington

As a battle, the engagement scarcely could be called a victory for the Allies, but Masséna had failed to relieve Almeida while Wellington had succeeded in covering its blockade. The total casualties of the British, Spanish, and Portuguese on the 3rd and 5th reached 1800, of the French nearly 3000, and 210 were taken prisoners. On the morning of the 8th May, the last of the enemy left the field, but three days later the Commander-in-Chief received bad news. On the previous night, the garrison of Almeida blew up part of the fortress and escaped, although the force sent by Wellington to blockade it was "four times more numerous than the garrison." He characterized it as "the most shocking event that has yet occurred to us." His correspondence at this period teems with references to it.

Masséna was no longer "the favoured child of victory" or Napoleon's "right arm," as the Emperor had called him. He was recalled, to be succeeded by Marmont, an excellent artillery officer then not quite thirty-seven years of age, whereas Masséna was fifty-three and deemed "too old" by his autocratic sovereign.

Marmont speedily came to the conclusion when he took up his new post that the so-called army of Portugal could not possibly expect to meet Wellington without rest and that there was not any likelihood of success. He accordingly moved

Duke of Wellington

his troops to the province of Salamanca, where we will leave them for a little while to watch the course of the war elsewhere.

Beresford had now invested Badajoz, and engaged the enemy in several sorties, on one occasion suffering severe loss owing to the imprudence of his troops. Receiving news to the effect that Soult was rapidly approaching with the solid object of relieving it, he raised the siege and posted his army on the ridge of Albuera to stop the French advance. The British Commander had nearly 32,000 men at his disposal. Of these no fewer than 24,000 were foreigners, including the Spanish forces of Blake, Castaños, Ballasteros, and Don Carlos d'España, which had formed a junction with him. The enemy had 23,000 troops.

As Wellington was not present a detailed description[153] of the battle, which took place on the 16th May, does not come within the province of this volume. It was one of the most fiercely contested of the entire war. So much so that Beresford used up his entire reserves and lost 4100 British troops, in addition to 1400 Portuguese and Spanish killed and wounded. The French losses were over 6000, and 500 were taken prisoners. Had it not been for Colonel Hardinge, Beresford would have retreated. Following his colleague's advice, he remained and was victorious. It was at Albuera that the

Duke of Wellington

57th Foot (now the 1st Middlesex Regiment) won the well-deserved name of "Die Hards" from the fact that Colonel Inglis shouted to his troops, "Die hard, my men; die hard!" **66** "It was observed," writes Beresford to Wellington, "that our sacrifice, particularly the 57th Regiment, were lying, as they had fought, in ranks, and that every wound was in front."

On the 19th Wellington, with two divisions, arrived at Elvas, and on the 21st he rode to Albuera and surveyed the site of the contest. "The fighting was desperate," he writes, "and the loss of the British has been very severe. The adverting to the nature of the contest, and the manner in which they held their ground against all the efforts the whole French army could make against them, notwithstanding all the losses which they had sustained. I think this action one of the most glorious and honourable to the character of the troops of any that has been fought during the war."

Surely a worthier tribute to the "regular" soldier was never penned!

CHAPTER XIV
The Siege of Ciudad Rodrigo (1811–12)

"The great object in all sieges is to gain time."

Wellington.

Outline of specific point of views or inside knowledge in this chapter:

67. He had recently received reinforcements from England.
68. Napoleon dominated practically the whole of Northern Europe. He was then planning a confederacy which was to consist of Sweden, Denmark, and the Grand Duchy of Warsaw.
69. Wellington's instructions to Hill will be found in "Dispatches," vol. viii. pp. 180–82.
70. "Cambridge Modern History," vol. ix. p. 469.
71. *i.e.* The province of Leon, in which Ciudad Rodrigo is situated.

The precise nature of the campaign was beginning to tell on Wellington. "I certainly feel, every day," he had written to the Earl of Liverpool on the 15th May 1811, "more and more the difficulty of the situation in which I am placed. I am obliged to go everywhere, and if absent from any operation, something goes wrong." "Another such battle" as Albuera, he informs his brother Henry on the 22nd, "would ruin us," and he proceeds to compare the Spanish and Portuguese troops, to the disadvantage of the former. They often held their ground too well, there was no moving them in a battle. On the other hand, "We do what we please now with the Portuguese troops; we maneuver them under fire equally with our own, and have some dependence on them; but these Spaniards can do nothing, but stand still, and we consider ourselves fortunate if they do not run away." In his report of the battle, Beresford mentions the Spanish cavalry as having behaved "extremely well."

Some idea of the enormous amount of labour involved may be gained from the fact that on the day mentioned Wellington either wrote or dictated at least eighteen dispatches. He including two dealing with the loss of an officer for whose widow and child he was endeavouring to obtain "favour and protection" at the hands of the home authorities. At the same time, he was actively preparing for the renewed siege of Badajoz: "The past action has made a serious hole in our ranks, but I am working hard to set all to rights again." He appeared "destined to pass his life in the harness," to use his own phrase, and had "an enormous quantity of business to settle of different descriptions."

Referring to the difference of opinion held by his officers regarding his policy, he says, "I believe nothing but something worse than firmness could have carried me through.... To this add that people in England were changing their opinions almost with the wind, and you will see that I had not much to look to excepting myself." The words are almost those of a broken-hearted man.

Badajoz

Was again invested on the 25th May, and the batteries opened fire on the 3rd of the following month in an attempt to breach the fort of San Christoval and the castle. Wellington had then made his headquarters at Quinta de Granicha. There he writes, on the 6th to the Earl of Liverpool, to the effect that if he cannot prevent the enemy from receiving provisions. He will not risk an action because he has not the means, and out of fairness to his soldiers he cannot "make them endure the labours of another siege at this advanced season. Notwithstanding that we have carried on our operations with such celerity," he concludes. "We have had great difficulties to contend with, and have been much delayed by the use of the old ordnance and equipment of Elvas, and of the Portuguese artillery, in this siege; some of the guns from which we fire are above 150 years old." The majority of them were supposed to be 24-pounders, but they proved to be larger, with the result that their fire was very uncertain. Two attempts were made to storm the outwork of San Christoval without success,156 many brave fellows perishing in the vain effort to escalate the walls.

Three weeks had not elapsed before it became eminently necessary to retire from this scene of activity. During this short time, nearly 500 officers and men had been reported as killed, wounded, or missing, and fifty-two of the Chasseurs Britanniques had deserted. "I have a great objection to foreigners in this army," he informs a colleague a little later, "as they desert terribly; and they not only give the enemy intelligence. Which he would find it difficult to get in any other manner but by their accounts and stories of the mode in which deserters from the French army are treated by us, some of them well-founded, they have almost put an end to desertion." The reason for the latter belief was the legend "that the deserters from the enemy are sent to the West India Islands and have no chance of ever returning to Europe."

Marmont, having united his scattered units, was about to join forces with Soult, which meant that when they marched on Badajoz, as undoubtedly they would do, the French army might number between 50,000 and 60,000 troops. Wellington had been of the opinion that it was possible to reduce the place before the end of the second week of June. An intercepted dispatch from Soult to Marmont made it abundantly evident that the enemy were to concentrate in Estremadura, and other intelligence clearly proved that the destination of the French army was "to the southward." Elvas, where supplies were running low, had first to be replenished so that it might be in a condition of defense should the enemy cross the frontiers. Leaving a comparatively small number of men to blockade Badajoz, and having made arrangements for the strengthening of Elvas, he marched from that place to Quinta de St João, where he remained for a considerable period. For nearly a fortnight, the French threatened to attack and had they done so it is scarcely possible that Wellington could have held his own in the field. Soult was the first to withdraw, the 157 immediate cause being the threatening of Seville by Blake, who retired when Soult approached. Marmont, feeling unequal to fight alone, marched to the valley of the Tagus and cantoned his army between Talavera and Plasencia. During the crisis, the two marshals mustered 62,000 troops, Wellington about 48,000.

The heat and other considerations prevented Wellington from besieging Badajoz; to relieve Cadiz was out of the question because the forces of Soult and Marmont would be almost assured to come to the assistance of the strength before the great southern port. He therefore, decided to besiege Ciudad Rodrigo, for four reasons stated in a letter to the Earl of Liverpool dated the 18th July. Namely he wrote: "We can derive some assistance from our militia in the north in carrying it into execution, and the climate in which the operation is to be carried on is not unfavorable at this season. If it should not succeed, the attempt will remove the war to the strongest frontier of Portugal; and, if obliged to resume the defensive, the strength of our army will be centrally situated, while the enemy's armies of the north and of the south will be disunited." Shortly after the above dispatch was written, he heard that Suchet had captured Tarragona, which made the proposed operation "less favorable." "However," he tells Beresford on the 20th of the same month, "we shall have a very fine army of little less than 60,000 men. **67** That will including artillery, in the course of about a fortnight. I do not see what I can do with it, to improve the situation of the allies, during the period in which it is probable that, the enemy's attention being taken up with the affairs of the north of Europe, we shall be more nearly on a par of strength with him, excepting we undertake this operation."

Lieutenant-General Hill

Was entrusted with the duty of watching the enemy in Alemtejo, 69 and two divisions were left in Estremadura. The Commander-in-Chief, with some 40,000 men, hastened towards Ciudad Rodrigo, unaware at the moment that the garrison had been reinforced and that Napoleon was sending more men to the Peninsula. When these important facts reached him, he contented himself with blockading the place and prepared to retire behind the Agueda should necessity warrant. Marmont sent for Dorsenne, who had taken the command in Galicia from Bessières, and with 60,000 troops set out toward the end of September to relieve Ciudad Rodrigo. Wellington then occupied El Bodon, on the left bank of the Agueda. "The object of taking a position so near to the enemy," he says, "was to force them to show their army. This was an object, because the people of the country, as usual, believed and reported that the enemy were not so strong as we knew them to be. If they had not seen the enemy's strength, they would have entertained a very unfavorable opinion of the British army, which it was desirable to avoid. This object was accomplished by the operations at the close of September."

Early on the morning of the 25th the Marshal drove in the outposts of Wellington's left wing and turned the heights occupied by the right center, thereby placing the British Commander in a dangerous position, from which he extricated himself by hurling his cavalry at the horsemen and artillery now endeavoring to scale the heights. Two British guns were captured and retaken at the point of the bayonet. When the French infantry were brought into action, Wellington gradually withdrew in the direction of Fuente Guinaldo, pursued by the enemy's cavalry, which were received by solid British squares and repelled as six miles were traversed. Marmont again advanced on the 26th but did not attack. Wellington159 retreated until he reached a permanent position in front of Sabugal on the 28th.

A rear-guard action had been fought on the previous day at Aldea da Ponte, but Marmont withdrew without offering battle, and, after supplying much-needed necessaries to Ciudad Rodrigo, proceeded to the Tagus valley and Dorsenne to Salamanca. Wellington renewed the blockade "in order," as he says, "to keep a large force of the enemy employed to observe our operations, and to prevent them from undertaking any job elsewhere." Placing his army in cantonments on the banks of the Coa, the Commander-in-Chief made his headquarters at Freneda.

While in their winter quarters both officers and men were able to recuperate after their previous tough campaign.

Duke of Wellington

Sports, theatricals and other amusements helped to pass away the time and to cheer up the army. Even more important was the opportunity thus afforded the many semi-invalids to recover their health. "We are really almost an army of convalescents." Wellington himself rode to hounds occasionally and applauded the amateur histrionic efforts of his soldiers when time and circumstances permitted him to attend their performances. He was able to re-establish Almeida as a military post, where he kept his battering-train to deceive the enemy, to blockade Ciudad Rodrigo, and to prepare for its investment.

Meanwhile, the guerillas were "increasing in numbers and boldness throughout the Peninsula," constantly annoying the French commanders. "It was their indomitable spirit of resistance," says Professor Oman, [70] "which enabled Wellington, with his small Anglo-Portuguese army, to keep the field against such largely superior numbers. No sooner had the French concentrated, and abandoned a district, then there sprang up in it a local Junta and a ragged apology for an army. Even where the invaders lay thickest, along the route 160 from Bayonne to Madrid, guerilla bands maintained themselves in the mountains, cut off couriers and escorts and often isolated one French army from another for weeks at a time. The greater partisan chiefs, such as Mina in Navarre,

Duke of Wellington

Julian Sanchez in Leon, and Porlier in the Cantabrian hills, kept whole brigades of the French in continuous employment. Often beaten, they were never destroyed, and always reappeared to strike some daring blow at the point where they were least expected. Half the French army was always employed in the fruitless task of guerilla-hunting. This was the secret which explains the fact that, with 300,000 men under arms, the invaders could never concentrate more than 70,000 to deal with Wellington."

In the autumn and winter of 1811 the enemy accomplished nothing of importance in eastern and southern Spain. In the south-east Suchet defeated Blake on the 25th October at the Battle of Sagunto, "the last pitched battle of the war," remarks the above authority, "in which a Spanish army, unaided by British troops, attempted to face the French." Forced into the city of Valencia with part of his motley array, Blake made a gallant attempt to rid himself of his besieger, an almost impossible task considering that Suchet had been reinforced while the unfortunate Spanish commander had been considerably reduced. On the 9th January 1812 his 16,000 followers laid down their weapons.

The investment of Ciudad Rodrigo by Wellington had been delayed owing to a complexity of causes. All the carting had to be performed by Portuguese and Spanish, and their slowness and the inclement weather combined precluded the Commander-in-Chief from pushing forward his operations with any celerity of movement. Empty carts took two days to go ten miles on a good road. Wellington confessed that he had to appear satisfied. Otherwise, the drivers would have deserted. If he succeeded in his designs he hoped to "make a solid campaign in the spring"; if he did not, "I shall bring back towards161 this frontier the whole [French] army which had marched towards Valencia and Aragon. By these means, I hope to save Valencia."

Alas, for human ambition! The capital of the province fell three days after the above dispatch was written.

On the 8th January 1812 a start was made, and Ciudad Rodrigo invested. During the night the palisaded redoubt on the hill of San Francisco, which the French had recently constructed, was stormed and carried, but Wellington at once perceived that the enemy had made good use of their time by strengthening their works and fortifying three convents in the suburbs. "The success of this operation," he writes, "enabled us immediately to break ground within 600 yards of the place, notwithstanding that the enemy still hold the fortified convents; and the enemy's work has been turned into a part of our first parallel and a noble communication made with it." Wellington encamped his men on the southern bank, which necessitated their fording the narrow stream, although he had built a bridge lower down the Agueda for munitions. It was no child's play for the soldiers. Through icy cold water, across ground covered with snow and frost, and amidst a rain of shot and shell, these brave fellows went to their work, each division in succession. Some of them returned, others did not, for "The path of glory leads but to the grave."

The convent of Santa Cruz was captured on the night of the 13th, followed on the 14th by the fall of the convent of San Francisco and other fortified posts in the suburbs. By these time batteries were within 180 yards of the walls. "We proceeded at Ciudad Rodrigo," he tells the Duke of Richmond, "on quite a new principle in sieges. The whole object of our fire was to lay open the walls. We had not one mortar; nor a howitzer. We are excepting to prevent the enemy from clearing the breaches, and for that purpose we had only two; and we fired upon the flanks and defences only when we wished to get the better of them, to (or "intending to") protecting those who were to storm. This shows the kind of place we had to attack...." Matters now became urgent, for advice had been received that Marmont was stirring. By the 19th, the breaches made in the ramparts by the artillery were declared practicable. Wellington had already summoned the Governor to surrender. His reply was that "he and the brave garrison which he commanded were prepared rather bury themselves in the ruins of a place entrusted to them by their Emperor." The troops, consisting of the regiments of the 3rd and Light Divisions and some Portuguese caçadores, marched to the assault in five columns. "Rangers of Connaught," cried General Picton to the "Fighting 3rd," who were charged with the center attack, "it is not my intention to expend any powder this evening; we'll do this business with the cold iron."

It was the task of Picton and his men to assault the great breach, while the 52nd, the 43rd, and the 95th Regiments, assisted by two battalions of caçadores, assaulted the other. At the same time, a brigade of Portuguese under General Pack was to make a feint at the Santiago gate, at the southern end of the town, and the light company of the 83rd regiment with another body of local soldiers were to scale the castle walls. As the columns advanced the moon, then in its first quarter, revealed their black outline to the enemy. They at once opened fire. No reply was vouchsafed by the Allies, who marched with fixed bayonets and unloaded muskets. It was not part of their plan to return a greeting made by men who were behind ramparts.

The Portuguese under Colonel O'Toole were the first to attack, closely followed by the 5th, 94th, and 77th regiments,

Duke of Wellington

the last supposed to act as a reserve. The Light Division, impatient of delay and not wishing to be rivalled in prowess, hurled themselves at the minor breach without waiting for the bags of hay which were to be thrown into the ditch to assist them in crossing. Many of the attacking force literally passed over the shot-riddled bodies of the vanguard as they attempted to get through. Major George Napier, while leading his men, had his arm shattered, but still continued to encourage them; Robert Craufurd, the intrepid and cantankerous commander of the Light Division, fell mortally wounded; Major-General Mackinnon was blown up by the explosion of a magazine. Nine officers and eleven non-commissioned officers and drummers gave up their lives for their country during the siege and in the assault from the 8th to the 19th, the total loss in killed and injured being nearly 1000. The hand-to-hand fighting continued in the streets, and the town caught fire.

At dawn 1700 of the enemy surrendered, including the Governor. Marmont's battering train, scores of field guns, and a plentiful supply of ammunition fell into the hands of the victors. Wellington had "great pleasure" in reporting "the uniform good conduct, and spirit of enterprise, and patience, and perseverance in the performance of excessive labor" on the part of the troops who had been engaged. As for the men themselves, they got drunk and sacked the place.

Wellington's rewards for the capture of Ciudad Rodrigo were numerous. He became an Earl in Great Britain, a Duke in Spain, and a Marquis in Portugal. Also, he was granted an extra annual pension of £2000 by Parliament. Financial offers were also forthcoming from the two Peninsula Powers, but he declined them. "He had only done his duty to his country, and to his country alone he would look for his reward."

Marmont was in ignorance of the siege until the 15th January. He then began to make preparations, but when he was ready the fortress had fallen, and he moved his army to Valladolid, to the north-east. Napoleon then sent orders to the Marshal that if he could not regain Ciudad Rodrigo he was to return to Salamanca, cross the frontier, and advance on Almeida. He foresaw that perhaps Wellington might turn his attention to Badajoz, which, in the Emperor's opinion, would be a "mistake,"164 and that of necessity he would have to return to succour the Portuguese fortress: "You will soon bring him back again." The British Commander also surmised that another attack on Ciudad was relatively possible. Before setting out on his next bold enterprise, he, therefore, put the fortifications in thorough repair and brought up a reserve supply of 50,000 rations in case it should be besieged. Satisfied that the place could now offer a heroic resistance to the enemy, and having also repaired the works of Almeida, he marched the greater part of his army to the valley of the Guadiana and invested Badajoz, which is on the left side of that river, on the 16th March 1812.

Wellington fully appreciated the immense value of time, and if he did not actually work with his eyes on the clock, he always endeavored to fix a definite date for his operations. Thus, as early as the preceding January he had written to his brother from Gallegos, a little to the north of Ciudad, that it was probable he would be in readiness to invest the place "in the second week in March." "We shall have full advantages in making the attack so early if the weather will allow it," he tells another correspondent. "First, all the torrents in this part of the country are then full, so that we may assemble nearly our whole army on the Guadiana, without risk to anything valuable here. **71** Secondly, it will be convenient to draw together our army at an early period in Estremadura, for the sake of the green forage, which comes in earlier to the south than here. Thirdly, we shall have advantages, in point of subsistence, over the enemy, at that season, which we should not have at a later period. Fourthly, their operations will necessarily be confined by the swelling of the rivers in that part as well as here." To deceive the enemy, he remained behind with the 5th Division as long as possible and gave instructions on a report to be circulated to the effect that he was going to hunt on the banks of the Huelbra and Yeltes.

CHAPTER XV
Badajoz and Salamanca (1812)

"I shall not give the thing up without good cause."

Wellington.

Outline of specific point of views or inside knowledge in this chapter:

72. Sir Harry Smith Autobiography, pp. 64–5.
73. Sir Herbert Maxwell, vol. i. p. 280.
74. A monument to the memory of Major-General Gaspard Le Marchant is in St Paul's Cathedral.

Considerable energy was displayed by the troops in the siege operations at Badajoz, notwithstanding the constant torrents of rain which soaked the men to the skin and filled the trenches as they worked. A bridge of pontoons was carried away and the flying bridges irretrievably injured by the swollen state of the Guadiana. The place was by no means an easy one to take, for strong outworks defended it, and Philippon, the French Governor, was the ablest officer in whom his troops placed every confidence. However, good fortune did not attend the first sortie made by about 2000 of the enemy on the 19th March. They were "almost immediately driven in, without affecting any object, with considerable loss, by Major-General Bowes, who commanded the guard in the trenches," to quote from Wellington's official dispatch.

On the 25th attack was made on Fort Picurina, a popular post separated from Badajoz by the little river called the Rivillas. Twenty-eight guns in six batteries were brought to bear upon it, and after dark the place was carried by storm, although it was protected by three rows of palisades defended by musketry. The garrison of the outwork consisted of about 250 men. Of these166 ninety, including the colonel, were taken prisoners, and most of the others were either killed or drowned in the swollen stream. An attempt was made to succour the brave defenders, but the soldiers were driven back before they could come up to the Picurina. The possession of this outwork enabled Wellington to place guns within 300 yards of the body of the place, and on the following day two breaching batteries began their work of destruction, with the result that on the 6th April three breaches were declared to be practicable.

At ten o'clock that night the attempt was to be made, the 3rd Division under Picton mounting the castle, the 4th Division with General the Hon. C. Colville attacking the bastion of La Trinidad, the Light Division commanded by Colonel Barnard the bastion of Santa Maria, General Leith's 5th Division the bastion of San Vincente. The attack on the bastions was to be made by storming the breaches. Wellington stood on rising ground facing the major breach, accompanied by the Prince of Orange and Lord March.

"When the head of the Light Division arrived at the ditch of the place (the great breach) it was a beautiful moonlight night," Sir Harry Smith relates with the authority of a participant in the action.[72] "Old Alister Cameron, who was in command of four Companies of the 95th Regiment, extended along the counterscarp to attract the enemy's fire. While the column planted their ladders and descended, came up to Barnard and said, 'Now my men are ready; shall I begin?' 'No, certainly not,' says Barnard. The breach and the works were full of the enemy, looking quietly at us, but not fifty yards off and most prepared, although *not firing a shot*. So soon as our ladders were already posted, and the columns

Duke of Wellington

in the very act to move and rush down the stepladders, Barnard called out, '*Now*, Cameron!' and the first shot from us brought down such a hail of fire as I shall never forget, nor ever saw before or since. It was[167] most murderous. We flew down the ladders and rushed at the breach, but we were broken, and carried no weight with us, although every soldier was a hero. The breach was covered by a breastwork from behind and ably defended on the top by *chevaux-de-frises* of sword-blades, sharp as razors, chained to the ground; while the ascent to the top of the breach was covered with planks with sharp nails in them.... One of the officers of the forlorn hope, lieutenant Taggart of the 43rd, was hanging on my arm—a mode we adopted to help each other up, for the ascent was most difficult and steep. A Rifleman stood among the sword-blades on the top of one of the *chevaux-de-frises*. We made a magnificent rush to follow, but, alas! in vain. He was knocked over. My old captain, O'Hare, who commanded the storming party, was killed. All were awfully wounded except, I do believe, myself and little Freer of the 43rd. I had been some seconds at the *revêtement* of the bastion near the breach, and my red-coat pockets were literally filled with chips of stones splintered by musket-balls. Those not knocked down were driven back by this hail of mortality to the ladders. At the foot of them I saw poor Colonel McLeod with his hands on his breast.... He said, 'Oh, Smith, I am mortally wounded.

Help me up the ladder.' I said, 'Oh, no, dear fellow!' 'I am,' he said; 'be quick!' I did so and came back again. Little Freer and I said, 'Let us throw down the ladders; the fellows shan't go out.' Some soldiers behind said, '... if you do we will bayonet you!' and we were literally forced up with the crowd. My sash had got loose, and one end of it was fast on the ladder, and the bayonet was very nearly applied, but the sash by pulling became free. So soon as we got on the glacis, up came a fresh Brigade of the Portuguese of the 4th Division. I never saw any soldiers behave with more pluck. Down into the ditch, we all went again, but the more we tried to get up, the more we were destroyed. The 4th Division followed us in marching up to the breach, and they made a most strange noise. The French saw us but took no notice.... Both Divisions were fairly beaten back; we never carried either breach (nominally there were two breaches) .... There is no battle, day or night, I would not willingly react except this. The murder of our gallant officers and soldiers is not to be believed."

The attack on the castle was no less furious. Again and again the ladders were hurled back, but they were always put in place again, notwithstanding the fearful and continuous fire to which the assailants were subjected. Great beams of timber, stones, everything calculated to kill or maim a man were regarded as useful weapons by the defenders. Nothing came amiss to them in their solid defense. Scores of soldiers

Duke of Wellington

were flung down when another minute of safety would have enabled them to secure a footing on the ramparts. They fell in the ditch, often injuring or killing others besides themselves. At last Lieutenant-Colonel Ridge managed to place two ladders at a spot which had not been used before, and where the wall was lower. The officer scaled one, followed by his men, and reached the rampart. The surprised garrison was repulsed, and very soon the castle was in the hands of the

British. Poor Ridge did not live to reap his richly-deserved reward. He was killed before the conclusion of the assault.

A little while previous to the successful termination of the attack Dr James McGregor and Dr Forbes approached Wellington. "His lordship," says the former, "was so intent on

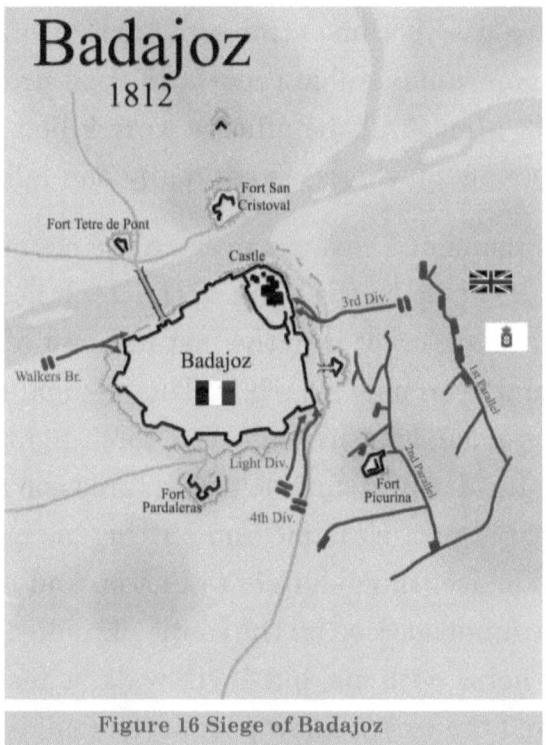

Figure 16 Siege of Badajoz

what was going on, that I believe he did not observe us. Soon

Duke of Wellington

after our arrival, an officer came up with an unfavorable report of the assault, announcing that Colonel McLeod and several officers were killed, with heaps of men who choked the approach to the breach. At the place where we stood, we were within hearing of the voices of the assailants and the assailed, and it was now painful to notice that the voices of our countrymen had grown fainter, while the French cry of '*Avancez, étrillons ces Anglais*,' became stronger. Another officer came up with still more unfavorable reports that no progress was being made, for almost all the officers were killed, and none left to lead the men, of whom a great many had fallen.

"At this moment I cast my eyes on the countenance of Lord Wellington, lit up by the glare of the torch held by Lord March. I never shall forget it to the last moment of my existence, and I could even now sketch it. The jaw had fallen, the face was of unusual length while the torch-light gave his countenance a lurid aspect, but still the expression of the face was fair. Suddenly turning to me and putting his hand on my arm, he said, 'Go over immediately to Picton, and tell him he must try if he cannot succeed on the castle.' I replied, 'My lord, I have not my horse with me, but I will walk as fast as I can. I think I can find the way; I know part of the road is swampy.' 'No, no,' he replied, 'I beg your pardon, I thought it was De

Lancey.' I repeated my offer, saying I was sure I could find the way, but he said 'No.'

"Another officer arrived, asking loudly, 'Where is Lord Wellington?' He came to announce that Picton was in the castle. He was desired instantly to go to the breach, and to request the Stormers to renew their efforts, announcing what had befallen; and immediately Lord Wellington called for his own horse, and followed by the Prince and Lord March, rode to the breach."

General Walker, leading the assault on San Vincente, experienced much the same rough treatment as the other divisions, but eventually succeeded in forcing his way into the town.

Philippon and a few hundred men managed to cross the Guadiana and found refuge in Fort San Christoval, only to surrender the following morning. The price paid by the victors in dead and wounded during the siege was nearly 5000 men; those of the enemy who laid down their arms numbered some 3800. The glory of the victorious army was unfortunately tarnished by the gross misconduct170 of the men, and it was not until some gallows was raised that a stop was put to their evil ways.

Duke of Wellington

Wellington was now anxious to meet Soult as soon as Badajoz was placed in a state of defense, but when he received the disturbing news of the defeat of the French garrison, the Marshal promptly retired to Seville. As Marmont was threatening Almeida and Ciudad Rodrigo the British General had no alternative but to turn northward. He had to thank the Spaniards for this. By neglecting to provision the places they had practically placed them at the mercy of the enemy, should he appear in considerable numbers. They were already blockading the latter place. "If Ciudad Rodrigo had been provisioned," Wellington writes to his brother Henry, "as I had a right to expect, there was nothing to prevent me from marching to Seville at the head of 40,000 men, the moment the siege of Badajoz was concluded." It was, of course, imperative that the line of communication between Marmont and Soult should be impeded as much as possible, and Hill was given this important task. Failing to surprise Almarez, the General pushed on to Fort Napoleon, on the other side of the Tagus, which was captured as well as Fort Ragusa. False information alone prevented Hill from following up his victories. He was told that Soult was in Estremadura, and he withdrew to the Guadiana.

Wellington began his advance, and on the 13th June crossed the Agueda. On his approach to Salamanca he at once laid siege to three newly-erected forts, "each defending the other." Marmont, knowing the likelihood of such an event, had wisely stored a good supply of food in them, so that there was a likelihood of their being able to hold out until he could succour them. The Marshal made one or two demonstrations to no good effect. He as obstinately declined to begin aggressive measures as Wellington dropped to be enticed to leave his solid position on the heights of San Christoval.

It took Wellington some time to secure the forts, which171 were well built and equipped, but on the 27th they fell into his hands, two by storm and one by capitulation. The last mentioned was being attacked when the flag was hauled down and would doubtless have been captured had not the commander given way before the British made good their assault. Marmont thereupon retired behind the Douro to await reinforcements.

After having destroyed some military works at Alba de Tormes and garrisoned the castle of Salamanca with Spaniards, Wellington pushed forward and engaged Marmont's rearguard on the 2nd July. He took up a position on the left bank of the Douro, on the opposite side of that occupied by the enemy, who was shortly afterwards strengthened by the support of Bonnet's division from the Asturias. Near Tordesillas, which with Toro and Tudela was held by the French, the Marshal took courage and fought an action with Sir Stapleton Cotton, who was in command of Wellington's right, on the 18th July. To resist him was impossible, for he had secured all the bridges and many of the fords. The action began at dawn, and Cotton gallantly maintained his post, but the enemy managed to turn the left flank of the British position. "The troops," says Wellington in his official report, "retired in admirable order to Torrecilla de la Orden, having the enemy's whole army on their flank, or in their rear, and thence to Guareña, which river they passed under the same circumstances, and effected their junction with the army."

Wellington fell back to within two miles of Salamanca, his left resting on the Tormes, his right abutting on two hills called Los Aripeles; [73] Marmont secured the heights of Nuestra Señora de la Peña.

Many years after the battle of Salamanca, General Alava, a Spanish officer at the British headquarters, was at Wellington's breakfast table at Walmer, and he regaled the company with the story of the great soldier's breakfast on the 22nd July 1812. Croker has recorded it for us,172 with the comment that the Duke listened "as quietly as if it related to another person."

Duke of Wellington

"The Duke had been very busy all the morning. He had not thought of breakfast, and the staff had continued starving. At last, however, there was a pause (I think he said about two) near a farmyard surrounded by a wall, where a kind of breakfast was spread on the ground, and the staff alighted and fell to; while they were eating, the Duke rode into the enclosure; he refused to alight, and advised them to make haste; he seemed anxious and on the look-out. At last they persuaded him to take a bit of bread and the leg of a cold roast fowl. Which he was eating without knife from his fingers, when suddenly they saw him throw the leg of the bird far away over his shoulder, and gallop out of the yard, calling to them to follow him. The fact is, he had been waiting to have the French *sighted* at a certain gap in the hills, and that was to be the signal of a long-meditated and long-suspended attack. 'I knew,' says Alava, with quiet drollery, 'that something *far-reaching was about to happen when an article so precious as the leg of a roast fowl was thus thrown away.*'" Croker adds that "the Duke sat by with his head inclined, quite silent, but with a quiet smile which seemed to say that the narration was a good deal pleasanter than the reality had been."

Wellington was able to seize the nearer hill, but the French secured the other while another small height named Nuestra Señora de la Peña was the center of a most desperate conflict, which continued through the long hours of the day.

Duke of Wellington

Marmont made the fatal error of dividing his army, sending Thomière's division to turn the British right flank, with intent to cut off all hope of retreat on the part of Wellington, should he wish to do so, using the Ciudad Rodrigo road. This movement separated the French left wing from the center, and this it was that caused the British Commander to fling away the dearly prized leg of a chicken.

After looking through his glass with wrapt attention Wellington turned to his Spanish colleague with the words, "My dear Alava, Marmont is undone!" His active brain told him at once of his enemy's mistake. Having made his dispositions, he ordered Pakenham, his brother-in-law, to throw the 3rd Division into line and cross the march of Thomière's columns. "It shall be done; give me your hand," replied that energetic officer. He hurled the Portuguese cavalry, two squadrons of the 14th Light Dragoons, and the "Fighting 3rd" at the flank and rear of the French left. Other divisions under Cole, Leith, Bradford, and Cotton, attacked the enemy in front. "No sooner was Pakenham in motion towards the heights," says one who took part in the battle, "than the ridge he was about to assail was crowned with twenty pieces of cannon, while in the rear of this battery were seen Foy's division, endeavoring to regain its place in the combat.

A flat space of 1000 yards in breadth was to be crossed before Pakenham could reach the height.

"The French batteries opened a heavy fire, while the two brigades of artillery, commanded by Captain Douglas, posted on a rising ground behind the 3rd Division, replied to them with much warmth. Pakenham's men may thus be said to have been between two fires, that of our own guns firing over their heads, while the French balls passed through their ranks, ploughing up the ground in every direction; but the veteran troops which composed the 3rd Division were not shaken even by this.

"Wallace's three regiments advanced in open columns, until within 250 yards of the ridge held by the French infantry. Foy's column, 5000 strong, had by this time reached their ground while in front of the hill had been hastily garnished with riflemen. All were impatient to engage, and the calm but the stern advance of Pakenham's right brigade was received with the beating of drums and loud cheers from the French. Whose light troops, hoping to take advantage of the time which the deploying into line would require, ran down the face of the hill in a state of great excitement. Pakenham instead, who was naturally of a boiling spirit and hasty temper, was on this day perfectly calm. He told Wallace to form a line from appropriate column without halting, and thus the different companies, by throwing forward their right shoulders, were in a line without the slow maneuver of deployment.

"Astonished at the rapidity of the movement, the French riflemen commenced an irregular and hurried fire, and even at this early stage of the battle a looker-on could, from the difference in the demeanor of the troops of the two nations, form a tolerably correct opinion of what the result would be. Regardless of the fire of the riflemen, and the showers of grape and canister, Pakenham continued to press forward. That way, his centre suffered, but still advanced; his right and left, being less oppressed by the weight of the fire, continued to advance at a more rapid pace, and as his wings inclined forward and outstripped the centre, his right brigade assumed the form of a crescent. The maneuver was a bold as well as a novel one, and the appearance of the brigade imposing and unique; because it so happened that all the British officers were in front of their men—a rare occurrence. The French officers were also in front, but their relative duties were widely different—the latter encouraging their men into the heat of the battle—the former keeping their true soldiers back—what a superb National contrast!"

When the brow of the hill was reached, the men were subjected to a murderous hail of fire from Foy's division. Nearly all of Wallace's first rank, as well as many officers, fell beneath it. But the others, urged by their commander, pressed on with fixed bayonets, and the French troops were forced backward. Thomière was amongst the killed, and many were taken prisoners in the rout which followed.

"Immediately on our left," the narrative continues, "the 5th Division were discharging vollies against the French 4th. Pack's brigade could be seen mounting the Aripeles height, but disregarding everything except the complete destruction of the column before him, Pakenham followed it with the brigade of Wallace, supported by the reserves of his division.

"The battle at this point would have been decided at the moment, had the heavy horse under Le Marchant been near enough to sustain him. The confusion of the enemy was so great that they became mixed pell-mell together, without any regard to order or regularity, and it was manifest that nothing short of a miracle could save Foy from total destruction. Sir Edward Pakenham continued to press on at the head of Wallace's brigade, but Foy's troops outran him. Had Le Marchant been aware of this state of the combat, or been near enough to profit by it, Pakenham would have settled the business by six o'clock instead of seven. An hour, at any period during a battle, is a serious loss of time, but in this action every moment was of vital import. The day was rapidly drawing to a close. The Tormes was close behind the army of

Duke of Wellington

Marmont, ruin stared him in the face. In a word, his left wing was doubled up—lost. Pakenham could have turned to the support of the 4th and 5th Divisions, had our cavalry been ready to back Wallace at the moment he pierced the column. This, beyond doubt, was the time by which to profit. The enemy might not have time to recollect himself. But then again, while Le Marchant was preparing to take part in the combat, Foy, with admirable presence of mind, remedied the terrible confusion of his division, and calling up a first brigade to his support, once more led his men into the fight, assumed the offensive, and Pakenham was now about to be assailed in turn. This was the most critical moment of the battle; Boyer's horsemen stood before us, inclining towards our right, which was flanked by two squadrons of the 14th Dragoons and two regiments of Portuguese cavalry; but we had a little dependence upon the Portuguese, and it behaved us to look to ourselves.

"Led on by the ardor of conquest, we had followed the column until at length we found ourselves in an open plain, intersected with cork trees, opposed by a multitude, who, reinforced, again rallied, and turned upon us with fury. Pakenham, Wallace, Seton, and Mackie, rode along the line from wing to wing, almost from rank to rank, and fulfilled the functions of adjutants in assisting the officers to reorganize the tellings-off of the men for square. Meanwhile, the first battalion of the 5th drove back some squadrons of Boyer's dragoons; the other six regiments were fast approaching the point held by Wallace, but the French cavalry in our front and upon our right flank caused Pakenham some uneasiness.

"The peals of musketry along the centre still continued without intermission. The smoke was so thick that nothing to our left was distinguishable. Some men of the 5th Division got intermingled with ours. The dry grass was set on fire by the numerous cartridge papers that strewed the field of battle. The air was scorching, and the smoke rolling on in huge volumes, nearly suffocated us.

"A loud cheering was heard in our rear—the brigade turned half round, supposing themselves about to be attacked by the French cavalry. A few seconds passed—the trampling of horse was heard—the smoke cleared away, and the heavy brigade of Le Marchant was seen coming forward in a line at a canter. 'Open right and left,' was an order quickly obeyed; the line untied, the cavalry passed through the intervals, and forming rapidly in our front prepared for their work.

"The French column, which a moment before held so imposing an attitude, became startled at this unexpected sight. A victorious and highly excited infantry pressed diligently upon them; a splendid brigade of three regiments of cavalry, ready to burst through their ill-arranged and beaten column, while no appearance of succour was at hand to protect them, was enough to appeal the boldest intrepidity. The plain was filled with the vast multitude; retreat was impossible, and the troopers came still pouring in, to join their comrades already prepared for the attack. It was too much for their nerves, and they sank under its influence, although they bravely made an effort to face the danger.

"Hastily, yet with much regularity, all things considered, they attempted to get into the square; but Le Marchant's brigade galloped forward before the evolution was half completed.

"The column hesitated, wavered, tottered, and then stood still! The motion of the many bayonets, as they clashed together, might be likened to a forest about to be assailed by a tempest, whose first warnings announce the ravage it is about to inflict. Foy's division vomited forth a terrific volley of fire as the horsemen thundered across the flat; Le Marchant was killed, [74] and fell down right in the midst of the French bayonets; but his brigade pierced through the vast mass, killing or trampling down all before them. The conflict was severe, and the troopers fell thick and fast, but their long, heavy swords, cut through bone as well as flesh....

"Such as got away from the sabres of the horsemen, sought safety among the ranks of our infantry, and scrambling under the horses. They ran to us for protection, like men who, having escaped the first shock of a wreck, will cling to any broken spar, no matter how little to depend on. Hundreds of beings frightfully disfigured, in whom the human face and form were almost obliterated—black with dust, worn down with fatigue, and covered with sabre cuts and blood—threw themselves among us for safety. Not a man was bayoneted—not even molested or plundered."

The battle still raged with unabated fury; "immediately in front of the 5th Division, Leith fell wounded as he led his men, but his division carried the point in dispute and drove the enemy before them up the hill.

"While these events were taking place on the right, the 4th Division, which formed the center of the army, met with a serious opposition. The more distant Aripeles, occupied by the French 122nd, whose numbers did not count more than 400, supported by a few pieces of cannon, was left to the Portuguese brigade of General Pack, amounting to 2000 bayonets. Falsely, though with well-founded reliance—their past conduct taken into the scale—Cole's division advanced into the plain, confident that all was right with Pack's troops, and a terrible struggle between them and Bonnet's corps took place. It was, however, but of short duration. Bonnet's troops were driven back in confusion, and up to this moment all had gone on well.

The three British Divisions engaged

Overthrew all obstacles, and the battle might be said to be won, had Pack's formidable brigade (formidable in numbers, at least) fulfilled their part. However, these men entirely failed in their effort to take the height occupied only by a few hundred Frenchmen, and thus gave the park of artillery that was posted on them, full liberty to turn its efforts against the rear and flank of Cole's soldiers. Nothing could be worse than the state in which the 4th Division was now placed, and the battle, which ought to have been and had been in a manner won, was still in doubt.

"Bonnet, seeing the turn which Pack's failure had wrought in his favor, re-formed his men, and advanced against Cole while the fire from the battery and small arms on the Aripeles height completed the confusion. Cole fell wounded; half of his division were cut off; the remainder in full retreat and Bonnet's troops pressing on in a compact body made it manifest that a material change had taken place in the battle and that ere it was gained some dirty up-hill work was yet to be done.

Figure 17 Battle of Salamanca Pakenham's Third Division

"Marshal Beresford, who arrived at this moment, galloped up to the head of a brigade of the 5th Division. He took out of the second line, and for a minute covered the retreat of Cole's troops. This force—composed of Portuguese—was insufficient to arrest the progress of the enemy, who advanced in the full confidence of an assured victory, and at this moment Beresford was carried off the field wounded. Bonnet's troops advanced, uttering loud cheers while the entire of Cole's division and Spry's brigade of Portuguese were routed. Our center was thus endangered. Boyer's dragoons, after the overthrow of the French left, countermarched, and moved rapidly to the support of Bonnet; they were also close in the track of his infantry, and the fate of this momentous

Duke of Wellington

battle might be said to hang by a hair. The fugitives of the 7th and 4th French Divisions ran to the succour of Bonnet. At the time they had joined him, his force had, indeed, assumed a formidable aspect, and thus reinforced it stood in an attitude far different from what it would have done, had Pack's brigade succeeded in its attack.

"Lord Wellington, who saw what had taken place by the failure of Pack's troops, ordered up the 6th Division to the support of the 4th, and the battle, although it was half-past 8 o'clock at night, recommenced with the same fury as at the outset.

"Clinton's division, consisting of 6000 bayonets, rapidly advanced to occupy its place in the combat. That division relieve the 4th from the awkward predicament in which it was placed, and essayed to gain what was lost by the failure of Pack's troops, in their feeble effort to wrest the Aripeles height from a few brave Frenchmen. They were received by Bonnet's troops at the point of the bayonet, and the fire opened against them seemed to be three-fold more heavy than that sustained by the 3rd and 5th Divisions. It was nearly dark, and the great glare of light caused by the thunder of the artillery, the continued blaze of musketry, and the burning grass, gave to the face of the hill a novel and terrific appearance. It was one vast sheet of flame, and Clinton's men looked as if they were attacking a burning mountain, the crater of which was defended by a barrier of shining steel. But nothing could stop the intrepid valor of the 6th Division as they advanced with a frantic resolution to carry the hill.

"The troops posted on the face of it to arrest their advance were trampled down and destroyed at the first charge, and each reserve sent forward to extricate them met with the same fate.

"Still Bonnet's reserves having attained their place in the fight, and the fugitives from Foy's division joining them at the moment, prolonged the battle until dark.

"These men, besmeared with blood, dust, and clay, half naked, and some carrying only broken weapons, fought with a fury not to be surpassed; but their impetuosity was at length calmed by the bayonets of Clinton's troops, and they no longer fought for victory, but for safety. After a desperate struggle, they were driven from their last hold in confusion, and a general and overwhelming charge, which the nature of the ground enabled Clinton's troops to make, carried this ill-formed mass of anxious soldiers before them, as a shattered wreck borne along by the force of some strong current. The mingled mass of fugitives fled to the woods and to the river for safety, and under cover of the night, succeeded in gaining the pass of Alba, over the Tormes. It was 10 o'clock at night—the battle was ended."

Marmont, who was wounded in the early part of the fight, lost 15,000 men, of whom 7000 surrendered to the British. The victors had nearly 700 officers and men killed, and over 4500 returned as wounded and missing. Six British Generals, including Wellington, whose thigh was grazed by a musket ball which had fortunately passed through his holster before it hit him, received injuries, and Le Marchant, as already mentioned, was shot. Of the enemy, four Generals were wounded and three killed, sufficient proof of the cruel nature of the long-continued contest. The victory would have been even more complete had the Spanish garrison at Alba de Tormes remained at their post instead of withdrawing without informing the Commander-in-Chief of their intention. As a consequence, the enemy was enabled to use the bridge there and make good their escape.

CHAPTER XVI
Closing of the Peninsular War (1812–14)

"In the whole of my experience I never saw an army so strongly posted as the French at the Battle of Toulouse. There ought to have been a concrete plan and description made of the whole affair as a matter of professional science."

Wellington.

Outline of specific point of views or inside knowledge in this chapter:

75. It is given in Gurwood, vol. x. pp. 61–66.
76. Lützen was fought on the 3rd May 1813, and Bautzen on the 20th and 21st May. In both battles the Prussians and Russians, who at the opening of the Leipzig campaign bore all the fighting for the Allies, were defeated. The only result of the armistice was that Austria threw in her lot with Russia, Prussia, and Sweden. —See the "Story of Napoleon," pp. 296–299.
77. "Personal Reminiscences of the Duke of Wellington by Francis, the first Earl of Ellesmere," p. 129. (London, 1903.)

Duke of Wellington

78. General Sir George T. Napier, pp. 255–260.

Marmont's army was not the only one in retreat. King Joseph, with 15,000 troops, had left Madrid with the certain purpose of joining the Marshal, but when he received news of the battle of Salamanca, he retreated on Valencia, where Suchet's army was posted, and peremptorily ordered Soult to evacuate Andalusia. This would enable him to bring 90,000 men to bear on his victorious enemy. His withdrawal from Madrid allowed Wellington to enter the capital on the 12th August 1812, Marmont, or rather Clausel, who had temporarily succeeded him, being driven back upon Burgos. The evacuation of the southern province was doubtless very gratifying to the Spaniards, but the threatened concentration of such a vast array of troops placed the Anglo-Portuguese army in an inferior position. The force at Wellington's disposal numbered 60,000 men, and although an additional 6000 had just landed at Alicante, in Valencia, it was evident that they would be of little service at the moment. When he became aware that Soult was about to abandon Andalusia he left part of his army to occupy Madrid, and with the remainder set out in the hope of being able to crush Clausel, who was at Valladolid. This he was unable to do, for the enemy retired from position to position. He followed him to Burgos, which Wellington entered, the French General meanwhile encamping on the banks of the Ebro, where he shortly afterwards received substantial reinforcements under Caffarelli and Souham the latter of whom arrived as Marmont's successor. Wellington was also joined by some 11,000 Spanish troops of the Army of Galicia. He at once laid siege to the castle above the town, which was toughly defended, and although the troops worked with praiseworthy ardor and four attempts were made to take it by assault, he was eventually forced to abandon the idea, and for a significant reason. Soult had joined King Joseph, and the combined army was on its way to Madrid. He had wasted a particular month, time which the French had used to full advantage.

It is related that one of the Irish regiments incurred his displeasure during the siege, and some of its members asked permission for it to lead one of the assaults. Their wish was granted, with the result that nearly all the men laid down their lives in the bold undertaking. When Wellington passed a little later, a soldier who had lost both his legs saluted and cried, "Arrah, maybe you are satisfied now, you hooky-nosed vagabond!" The Commander could not restrain a smile, and promptly sent assistance. The Irishman ended his days in Chelsea Hospital.

Duke of Wellington

Sending word to Hill to abandon Madrid and meet him on the Tormes, Wellington skillfully withdrew his men from Burgos, and although his rear-guard was much harassed by Souham's troops, he formed a junction with his lieutenant near the battlefield of Salamanca. On arrival on the Tormes, they were almost face to face with the allied army, but divided counsels reigned, and he skillfully eluded the French, although they turned his position. Aided by a dense fog, Wellington managed to slip away unperceived. After a sharp engagement at a ford of the Huebra, the pursuers abandoned the attempt to secure the roads to Ciudad Rodrigo, which place was reached by Wellington on the 18th November. Soult retired to Toledo, Souham to Valladolid and Joseph to Segovia.

A pen-sketch of the men during this terrible retreat tells us that "such a set of scarecrows never was seen. It was difficult to say what they were, as the men's coats were patched with grey, some had blankets over them, and most were barefooted; every step they took was up to the knees in mud; women and sick men were actually sticking in it.... A brigade of cavalry, however, which was covering the rear, had left Lisbon but a short time before, and was in high order. The clothing of the men scarcely soiled, and the horses sleek and fat, made a strange contrast with the others, especially the company of artillery that had served in the batteries before Burgos. We at first took the latter for prisoners, as they were mostly in French clothing, many of them riding in the carriages with the sick and wounded, drawn, some by oxen, and some by mules and horses. I never saw British soldiers in such a state."

Wellington and his men then went into cantonments, the former making his headquarters at Freneda. Much was done to improve the *morale* of the troops, who had got into a very insubordinate state. Reinforcements came to hand, and Wellington worked hard to reorganize the Spanish army, of which he had been appointed Generalissimo after the battle of Salamanca. He had also been raised to the rank of Marquis, thanked by both Houses of Parliament, and presented with £100,000. He paid a visit to the Cortes, made a speech, and wrote a long letter to one of the Deputies. In that letter, he criticized "the powers that be" in no uncertain way, adding, however, some measures which would "give your Government some chance of standing, and your country some chance of avoiding further revolutions." The full communication must be studied to be fully appreciated. 75 "The Government and the Assembly," he says in one passage, "instead of drawing together, are like two independent powers, jealous and afraid of each other; and the consequence is, that the machine of

Duke of Wellington

Government is at a stand. To this add that the whole system is governed by little local views, as propounded by the daily press of Cadiz, of all others the least enlightened and the most licentious." "I will fight for Spain as long as she is the enemy of France, whatever may be her system of government," he adds, "but I cannot avoid seeing and lamenting the evils which await the country if you do not retrace your steps, let what will be the result of the military operations of the war...."

He advised the establishment of a permanent Regency, "with all the powers allotted by the Constitution to the King, in the hands of one person." He, or she, should be aided by a Council, whose five members should superintend the Department de Estado, the Interior and Ultramar, Gracia y Justicia, Hacienda, and of War and of Marine respectively, each being responsible for the department under his superintendence. He suggested either turning "the Council of State into a House of Lords," or making "a House of Lords of the Grandees, giving then concurrent powers of legislation with the Cortes; and you should leave the patronage now in the hands of the Council of State in the hands of the Crown."

In these days of Socialism, the following remarks, which occur in the same letter, are of more than passing interest. "The theory of all legislation," he says, "is founded in justice; and, if we could be positive that legislative assemblies would on all occasions act according to the principles of justice, there would be no event for those checks and guards which we have seen established under the best systems. Unfortunately, however, we have seen that legislative assemblies are swayed by the fears and passions of individuals; when unchecked, they are tyrannical and unjust; nay, more: it, unfortunately, happens too frequently that the most dictatorial and unfair measures are the most popular. Those measures are mostly widely held which deprive rich and powerful individuals of their properties under the pretense of the public advantage; and I tremble for a country in which, as in Spain, there is no barrier for the preservation of private property, excepting the justice of a legislative assembly possessing supreme powers."

In summing up the result of his operations in the field during 1812, Wellington tells the Earl of Liverpool on the 23rd November, that notwithstanding adverse criticism in the newspapers, "it is in fact the most successful campaign in all its circumstances, and has produced for the cause more significant results than any campaign in which a British army has been engaged for the last century. We have taken by siege Ciudad Rodrigo, Badajoz, and Salamanca; and the Retiro surrendered. In the meantime, the Allies have taken Astorga, Guadalaxara, and Consuegra, besides other places occupied by Duran and Sir H. Popham. In the months elapsed

Duke of Wellington

since January, this army has sent to England little short of 20,000 prisoners. They have taken and destroyed or have themselves the use of the enemy's arsenals in Ciudad Rodrigo, Badajoz, Salamanca, Valladolid, Madrid, Astorga, Seville, the line before Cadiz, etc. Upon the whole we have taken and destroyed, or we now possess, little short of 3000 pieces of artillery. The siege of Cadiz has been raised, and all the countries south of the Tagus have been cleared of the enemy.

"We should have retained still greater advantages, I think, and should have remained in possession of Castile and Madrid during the winter, if I could have taken Burgos, as I ought early in October, or if Ballasteros had moved upon Alcarez as he was ordered, instead of intriguing for his own aggrandizement.

"The fault of which I was guilty in the expedition to Burgos was, not that I undertook the operation with inadequate means, but that I took there the most inexperienced instead of the best troops.... I see that a disposition already exists to blame the Government for the failure of the siege of Burgos. The Government had nothing to say to the siege. It was entirely my own act. Regarding means, there were ample means both at Madrid and at Santander for the siege of the strongest fortress. That which was wanting at both places was means of transporting ordnance and military stores to the place where it was desirable to use them.

"The people of England, so happy as they are in every respect, so rich in resources of every description, having the use of such excellent roads, will not readily believe that imperative results here frequently depend upon 50 or 60 mules more or less, or a few bundles of straw to feed them. However the fact is so, notwithstanding their incredulity...."

When Wellington was ready for his 1813 campaign he had 75,000 British and Portuguese at his disposal, and some 60,000 Spaniards, in addition to the irregular bands which were the bane of the enemy. The different French armies totaled some 200,000 troops, but it was deemed necessary to send 40,000 of these, under Clausel and Foy, to exterminate the *guerilleros*, which was to Wellington's advantage, especially as it was impossible for Napoleon, now deeply involved owing to the disastrous Russian campaign, to send further reinforcements. Soult was withdrawn, with 20,000 men, to oppose the Russian advance. By way of further encouragement, Andalusia, Estremadura, Galacia, and the Asturias no longer sheltered the enemy. The British left was under Graham, the right under Hill, and the center under the Commander-in-Chief. The first marched upon Valladolid, the French retreating before him, and was joined near Zamora on

Duke of Wellington

the 1st June 1813 by Wellington, followed two days later by Hill. The French were deceived by these movements, for they expected the main attack to be made from Ciudad Rodrigo and Almeida with the object of occupying Madrid. This was far from Wellington's purpose, which was to carry on the war in the northern provinces, sever the French communications with the homeland, and force them to withdraw to the Pyrenees. King Joseph hastily retired from Valladolid and reached Burgos. On the approach of Wellington to that town, the fortifications were blown up and the enemy fell back beyond the Ebro.

"When I heard and saw this explosion (for I was within a few miles, and the effect was tremendous)," Wellington remarks, "I made a rapid resolution forthwith—more instant to cross the Ebro, and endeavor to push the French to the Pyrenees. We had heard of the battles of Lützen and Bautzen and of the armistice, [76] and the affairs of the Allies looked very ill. Some of my officers remonstrated with me about the impudence of crossing the Ebro and advised me to take up the line of the Ebro, etc. I asked them what they meant by taking up the line of the Ebro, a river 300 miles long, and what good I was to do along that line? In short, I would not listen to the advice; and that very evening (or the very next morning) I crossed the river and pushed the French till I afterwards beat them at Vittoria."

"We continued to advance," writes a soldier of the 71st Regiment who fought in the battle, "until the 20th of June; when reaching the neighborhood of Vittoria, we encamped upon the face of a hill. Provisions were very scarce. We had not a bit of tobacco, and were smoking leaves and herbs. Colonel Cadogan rode away and got us half a pound of tobacco a man, which was most welcome.

"Next morning we got up as usual. The first pipes played for the parade; the second did not play at the typical time. We began to suspect all was not right. We remained thus until eleven o'clock; then received orders to fall in, and follow the line of march. During our march, we fell to one side, to allow a brigade of guns to pass us at full speed. 'Now,' said my comrades, 'we will have work to do before night.' We crossed a river, and, as we passed through a village, we saw, on the other side of the road, the French camp, and their fires still burning, just as they had left them. Not a shot had been fired at this time. We observed a hefty Spanish column moving along the heights on our right. We halted, and drew up in the column. Orders were given to brush out our locks, oil them, and examine our flints. We being in the rear, these were soon followed by orders to open out from the center, to allow the 71st to advance. Forward we moved up the hill. The firing was now hefty. Our rear had not engaged before word came for the doctor to assist Colonel Cadogan, who was wounded. Immediately we charged up the hill, the piper playing, 'Hey Johnny Cope.' The French had possession of the top, but we soon forced them back and drew up in the column to the height; sending out four companies to our left to skirmish. The remainder moved on to the counter height. As we advanced driving them before us, a French officer, an appealing fellow, was pricking and forcing his men to stand. They heeded him not—he was very harsh. 'Down with him!' cried one near me; and down he fell, pierced by more than one ball.

"Scarce were we upon the height, when a heavy column, dressed in great-coats, with white covers on their hats, exactly resembling the Spanish, gave us a volley, which put us on the right about at double-quick time down the hill, the French close behind, through the whines. The four companies got the word the French were on them. They likewise thought them Spaniards, until they got a volley that killed or wounded almost every one of them. We retired to the height, covered by the 50th, who gave the pursuing column a volley which checked their speed. We moved up the remains of our shattered regiment to the height. Being in great want of ammunition, we were again served with sixty rounds a man, and kept up our fire for some time, until the bugle sounded to cease firing....

"At this time the Major had the command, our second Colonel being wounded. There were not 300 of us on the height able to do duty, out of above 1000 who drew rations in the morning. The cries of the wounded were most heart-rending.

"The French, on the counter height, were getting under arms; we could give no assistance as the enemy appeared to be six to one of us. Our orders were to maintain the height while there was a man of us. The word was given to shoulder arms. The French, at the same moment, got under arms. The engagement began on the plains. The French were amazed and soon put to the right about, through Vittoria. We followed, as quick as our weary limbs could carry us. Our legs were full of thorns, and our feet bruised upon the roots of the trees. Coming to a bean field at the bottom of the heights, immediately the column was broken, and every man filled his haversack. We continued to advance until it was dark and then encamped on a height above Vittoria.... I had fired 108 rounds this day."

According to the official figures, the British lost 740 men by death and 4174 were wounded, out of a total strength of 80,000. The captures included 151 guns, 415 caissons,

Duke of Wellington

14,249 rounds of ammunition, nearly 2,000,000 musket ball cartridges, 40,668 lbs. of gunpowder, fifty-six forage wagons, forty-four forge waggons, a treasure to the amount of £1,000,000, pictures by Velasquez and other masters, jewelry, public and private baggage. King Joseph's carriage, and Jourdan's baton. The last-mentioned was given by Wellington to the Prince Regent, who with becoming fitness sent the donor a Field-Marshal's baton. The French had 65,000 men engaged in the battle of Vittoria, of whom some 6000 were killed and wounded, and 1000 taken prisoners.

The defeated army crossed the Pyrenees and marched to Bayonne, where it was joined by the troops under Foy and Clausel, who had been pursued by the Allies. "To hustle the French out of Spain before they were reinforced," had been Wellington's object, and he had carried it out completely. As the garrisons of the fortresses of Pampeluna and San Sebastian had been strengthened, the former by Joseph and the latter by Foy, during their retreat, Wellington now turned his attention to them. Although the army under Suchet was the only one now left in the Peninsula, it occupied Catalonia and part of Valencia, and might, therefore, attack Wellington's right flank.

Napoleon was at Dresden

When he heard of his brother's disaster at Vittoria, and he was in no mood for soft words. He recalled both Joseph and Jourdan and gave the command to Soult. "It is hard to imagine anything so inconceivable as what is now going on in Spain," the Emperor writes to Savary on the 3rd July 1813. "The King could have collected 100,000 picked men: they might have beaten the whole of England." He blamed himself for the "mistaken consideration" he had shown his brother, "who not only does not know how to command but does not even know his own value enough to leave the military command alone."

Duke of Wellington

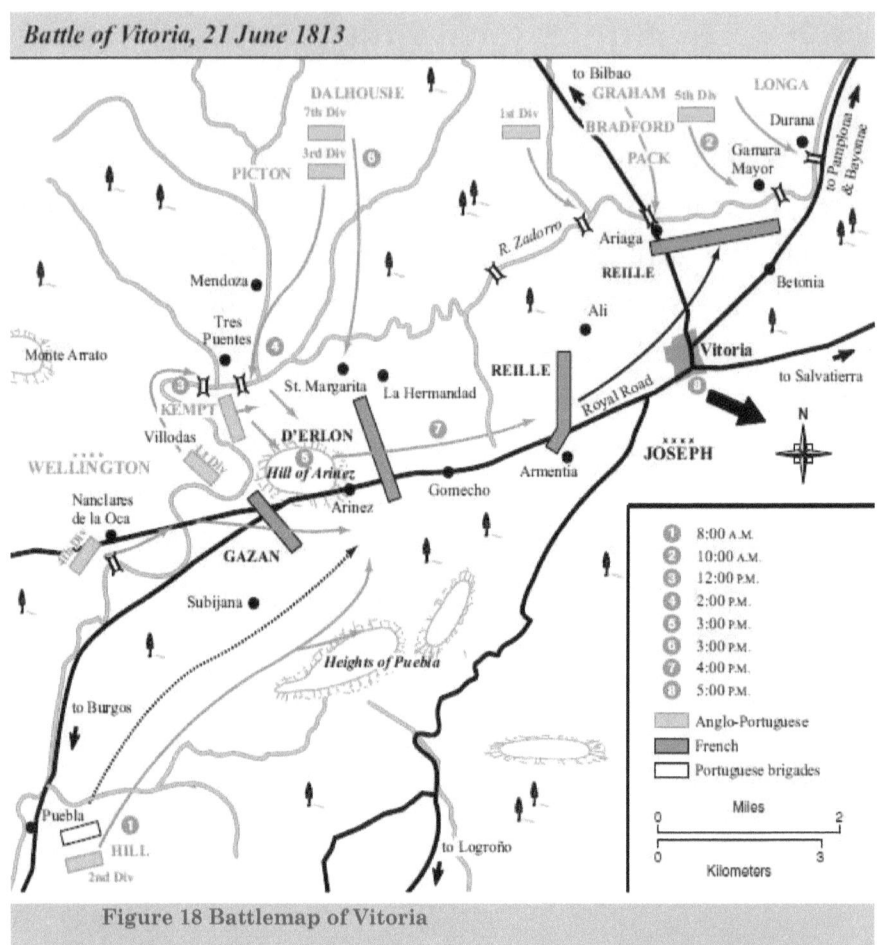

Figure 18 Battlemap of Vitoria

Soult reached Bayonne on the 12th July, and thirteen days later had marched on Pampeluna with 73,000 troops, bent on relieving one or other of the fortresses, perhaps both. He attacked the British right at Roncesvalles and turned the position; Hill was attacked at the head of the valley of Baztan and was obliged to withdraw. Wellington at once raised the siege of San Sebastian, which had been carried on by Sir Thomas Graham, and contented himself with blockading the fortress. He immediately concentrated his right and center at Sorauren, near Pampeluna. The series of fights which took place at this time is known as the battles of the Pyrenees. On the 27th Wellington arrived, and a rousing cheer greeted him, which it is said deterred the French from making anything but unfair attacks. Probably the truth of the matter is that Soult hesitated because he was expecting additional forces with d'Erlon, for the Marshal was scarcely likely to be overawed by a greeting. A corporal, unable to restrain his enthusiasm as the Commander-in-Chief rode along the line, shouted out to the intense amusement of all, "There goes the little blackguard what whops the French!" 77

Soult was pointed out to the General by a spy. "Yonder," Wellington is reported to have said, "is a great but cautious general; he will delay his attack to know the reason of those cheers; that will give time for my reinforcements to come up, and I shall beat him." As a matter of fact, the 6th Division of infantry, to which Wellington had referred, did arrive, and "bludgeon work," to use his expression, took place on the 28th, the anniversary of Talavera. The reinforcements had scarcely secured their position, their right resting on Orcain and their left on the heights overlooking the valley of the Lanz, then a very determined attack was made by the enemy. They were driven back and made an attempt on the hill occupied by the 7th caçadores and Ross's brigade of the 4th Division. They obtained possession of it for a short time until driven down. When the battle became general, the 10th Portuguese regiment was overpowered, necessitating the withdrawal of Ross. Wellington then ordered two regiments to charge the enemy on the heights and those on the left, with the result that the French were "driven down with immense loss." "Every regiment," says Wellington, "charged with the bayonet, and the 40th, 7th, 20th, and 23rd, four different times." Of Wellington's 16,000 troops he lost 2600 killed and wounded, French 1800 out of 20,000. The Portuguese behaved "admirably," and the Commander-in-Chief "had every reason to be satisfied with the conduct of the Spanish Regiments El Principe and Pravia."

By sunset, Soult's attacks had diminished, and on the following morning he began to retreat, although he received reinforcements to the number of 18,000 troops. On the 30th, he attacked Hill to no good effect, and Wellington forced the French to retire from a strong position they had taken up. The pursuit continued until the 1st August when it was discontinued, for the Allies were in possession of the passes and the strenuous exertion of the troops was beginning to tell upon them.

Wellington again took up his headquarters at Lesaca. Writing to Graham, he says, "I hope that Soult will not feel any inclination to renew his expedition. The French army must have suffered greatly. Between the 25th of last month and 2nd of this, they were engaged seriously not less than ten times; on many occasions in attacking fixed positions, in others beat them or pursued. I understand that their officers say that they have lost 15,000 men. I thought so; but as *they* say so, I now think *more*. It is strange enough that our diminution of strength to the 31st does not exceed 1500 men, although, I believe, our casualties are 6000." It was on the 31st that San Sebastian fell, the castle capitulating shortly afterwards, and the day is also noteworthy for Soult's attack on San Marcial, which was repulsed by Spanish troops, the enemy retiring across the Bidassoa. Unfortunately, Sir John Murray made no headway against Suchet in the east of Spain and was superseded by Lord William Bentinck, who besieged Tarragona, which his predecessor had evacuated. Although he was compelled to retire on the approach of the French Marshal, the city was eventually occupied by the British troops. Their entry into Villa Franca was marred by the rout of the advanced

Duke of Wellington

guard in the pass of Ordal, necessitating their retreat towards Tarragona.

It is evidently impossible to unravel with any approach to detail the tangled skein of various maneuvers which took place at this period, perhaps the most trying and exacting of the war in the Peninsula. Gleig, however, gives one charming touch to a complex picture which reveals more of the personality of the great General than many pages of military movements and is infinitely more valuable for the purposes of a life story. "Lord Wellington," he records, "after directing a Spanish column to move up a glen towards a particular point, looked at his watch, and observed to those about him that it would take the men so much time to perform the journey. He added that he was tired, and dismounting from his horse, wrapped himself in his cloak and went to sleep. A crowd of officers stood around him, and among others some Spanish generals, whose astonishment at the coolness of their chief was expressed in audible whispers. For the very crisis of the struggle was impending, and the French being in greater strength upon the spot seemed to have the ball at their foot. Now, among the officers of the headquarters' staff, there were several who had never approved the passage of the Ebro. These began to speak their minds freely, and one, the bravest of the brave, the gallant Colonel Gordon, exclaimed, 'I always thought it would come to this. I was sure we should make a mess of it if we got entangled among the Pyrenees, and now see if my words don't come spot-on.' Lord Wellington happened to awake just as Gordon thus unburdened his conscience. He sat up, and without addressing himself to anyone, in particular, extended his right hand open, and said, as he closed it, 'I have them all in my hand, just like that.' Not another word was spoken. The Spaniards had reached the top of the glen; Lord Wellington and his attendants remounted their horses, and the battle was renewed."

On the 7th October 1813 Wellington passed the Bidassoa to the left of his army. Soult was attacked and driven back with the loss of eight pieces of cannon, taken by the Allies in the captured redoubts and batteries. The fighting was continued on the following day, after the fog which obscured the enemy's position had lifted. When a rock occupied by the French to the right of their position was carried "in the most gallant style" by the Spaniards, who immediately afterwards distinguished themselves by transport an entrenchment on a hill which protected the right of the camp of Sarre. Soult withdrew during the following night, and took up a series of entrenched positions behind the Nivelle, leaving, as Alison so eloquently puts it, "a vast hostile army, for the first time since the Revolution, permanently encamped within the territory of France. And thus was England, which throughout the contest had been the most persevering and resolute of all the opponents of the Revolution. Whose government had never yet either yielded to the victories or acknowledged the Chiefs which it had placed at the head of affairs, the, first of all, the forces of Europe who succeeded in planting its victorious standards on the soil of France."

On the 10th November, a little over a week after the surrender of Pampeluna through starvation, for the fall of which he had waited before resuming offensive operations. Wellington, with an army of about 90,000 men, attacked the enemy's position, an exceedingly resilient one, the right extending from the sea to St Jean de Luz, the left from Bidarray to St Jean Pied de Port, the centre between Amotz and Ascain. The enemy were driven out of the lines and followed the river, with a loss of 4200 men and no fewer than 51 guns, the Allies losing about 2500 killed and wounded.

"Our loss," says Wellington, "although severe, has not been so great as might have been expected, considering the strength of the positions attacked, and the length of time, from daylight in the morning till night, during which the troops were engaged." The disorders which followed the battle were so great that except a single division Wellington sent the whole of the Spaniards—some 25,000—back to the Peninsula.

Duke of Wellington

On the 12th, the Marshal was in the entrenched camp at Bayonne, and six days later the victorious army went into cantonments, where it remained until the 9th December when it was ordered to march towards Bayonne. It forced the passage of the Nive, and a series of engagements was fought until the 13th, on which date Hill, with one British and one Portuguese division, fought and won the battle of Saint Pierre. Wellington came up but refrained from interfering, and when he saw that his brave colleague had proved victorious, he wrung his hand in a hearty grip and exclaimed, "Hill, the day is entirely your own."

In the battles of the Nive

The French lost 5800 killed and wounded, and three German regiments by desertion to the Allies, whose losses totaled 4600. Soult had now one of two alternatives, either to be hemmed in at Bayonne or to retreat. He chose the latter, and marched in the direction of Toulouse with less than 40,000 troops, Napoleon, now in a desperate plight, having withdrawn 10,000 for the defense of the eastern frontier of France.

Leaving Sir John Hope to blockade Bayonne, Wellington followed Soult, who took up a position at Orthez, on the right bank of the Gave de Pau. Early on the morning of the 27th February the battle opened by Beresford turning the enemy's right, but he was driven back, as was Picton, who attacked the enemy's center. *"Enfin je le tiens!* —At last I have him!" exclaimed Soult, but Wellington changed his plan, and at once sent Hill to cross the river by the ford above Souars and cut off the Marshal's retreat by the great road to Pau. At the same time, he ordered two divisions against the right of the enemy's center, and Colborne cut off the division which had checkmated Beresford. The French under Reille were driven from the heights, and at first retired, in good order, but Cotton and Lord Edward Somerset charged and spread considerable confusion in the ranks while Hill marched on Aire and attacked Clausel. The Portuguese were repulsed, but the British drove the enemy from the town with an excessive loss.

Wellington was wounded almost at the end of the battle, which is perhaps one reason why the pursuit was not so rapid as it might otherwise have been. However, Beresford was sent with two divisions to Bordeaux, whose citizens bade them enter, and thereupon proclaimed the Duc d'Angoulême, eldest nephew of Louis XVIII, who was now with the British army, as Prince Regent.

The last battle of the Peninsular War

was fought on Easter Sunday, the 10th April 1814, at Toulouse, on which Soult's army had concentrated.

Duke of Wellington

A mistake on the part of an engineer as to the breadth of the Garonne above Toulouse prevented Wellington from crossing at the spot he had selected because there were not sufficient pontoons. This caused considerable delay and a march to a narrower but more difficult place below the town. Sir George Napier says that he never saw the Commander-in-Chief in such a rage—he was "furious." On the completion of the gangway, Beresford, with a portion of the army, passed over, drove in the French outposts, and remained in front of the enemy. There they stopped for three days, cut off from the main force and liable to attack any moment. This unexpected situation was brought about by a storm which flooded the river and swept away the pontoons.

Soult is stated to have given this reason for failing to assail Beresford's force: "You do not know what stuff two British divisions are made of; they would not be conquered as long as there was a man of them left to stand, and I cannot afford to lose men now."

When the new bridge was available no time was lost in crossing the river, and on the 10th Soult was attacked. An eye-witness thus records the event [78]:

"The 4th, 6th, and a Portuguese division under Marshal Beresford's orders attacked the great fort on the

Duke of Wellington

right of the French. There was the brunt of the battle, for the enemy was strongly posted and flanked by works, with trenches in their front, and their best troops opposed to ours. But nothing could damp the courage of this column; the enemy's guns poured a torrent of fire upon it; still it moved onward. As column upon column appeared, crowning the hill and forming lines in front and on the flanks of our brave fellows who were near the top. Then such a roll of musketry accompanied by peals of cannon and the shouts of the enemy commenced, that our soldiers were honestly forced to give way and were driven down again. This attack was twice renewed and twice were our gallant fellows forced to retire. As they being got into order again and under a tremendous fire of all arms from the enemy. They once more marched onward determined '*to do or die*' (for they were nearly all Scotch) and, having gained the summit of the position, they charged with the bayonet, and in spite of every effort of the enemy, drove all before them and entered every redoubt and fort with such a courage as I never saw before. The enemy lay in *heaps*, dead and dying! Few, very few, escaped the slaughter of that day; but 'victory' was heard shouted from post to post as that gallant band moved along the crown of the enemy's position taking every work at the point of the bayonet.

Duke of Wellington

"While the work of death was going on here, the center of the French position was attacked by the Spanish column of 8000 men, under General Freyre, who had *demanded* in rather a haughty tone that Lord Wellington should give the Spaniards the post of honor in the battle. He acceded but took special care to have the Light Division in reserve to support them in case of *accidents*. Old Freyre placed himself at the head of his column, surrounded by his staff, and marched boldly up the hollow way, or road, which led right up to the enemy, under a hefty and destructive fire of cannon shot. These, which plunging into the head of his column made great havoc among his men; still they went steadily and boldly on, to my astonishment and delight to see them behave so gallantly, and I could not help expressing my happiness to Colonel Colborne. But, alas! He knew them too well, and said to me, 'Gently, my friend; don't praise them too soon; look at yonder brigade of French Light Infantry, ready to attack them as soon as the head of their column enters the open ground. One moment more and we shall see the Spaniards fly! Gallop off, you, and throw the 52nd Regiment (which was in line) into open column of companies, and let these fellows pass through, or they will carry the regiment off with them.' He had scarcely finished the words when a well-directed fire from the French Infantry opened upon the Spanish column, and instantly the words '*Vive l'Empereur! En avant! en avant!*' accompanied by

Duke of Wellington

a charge, put the Spaniards to flight, and down they came upon the 52nd Regiment. I had but just time to throw it into an open column of companies when they rushed through the intervals like a torrent and never stopped till they arrived at the river some miles in the rear. As soon as they had passed, and I had formed the regiment into line again, we moved up and took the Spaniards' place. We were driving before us the enemy's brigade, who, being by this time entirely beaten on the right and all his forts and trenches carried by Beresford's troops, had retreated into the town. So that we found the fort on that part of the position which we attacked quite abandoned, and we entered it without loss.

"On our right the 3rd Division, under General Picton, was ordered to make a false attack on the canal bridge. That bridge, which was strongly fortified. It formed an

Figure 19 Battle of the Pyrenees

impracticable barrier to that part of the town; but General Picton (who never hesitated at disobeying his orders) thought proper to change this false attack into a real one, and after

Duke of Wellington

repeated and useless attempts to carry it was forced to give it up, with an immense loss of officers and men. To our extreme right and on the opposite side of the river General Hill was stationed with his corps to watch the bridge and gates of the town. In either prevent any attempt of the enemy to pass over a body of troops during the action to cut off our communications with the rear or, should he show any design of retreating that way, to impede him. However, all was quiet on that side, and now that every man of the enemy's army had been chased from the position the battle was won, and the roar of cannon, the fire of the musketry, and the shouts of the victors ceased. All was still; the pickets placed; the sentinels set, and the greatest part of the army sleeping in groups around the fires of the bivouac."

Soult had only been able to bring some 39,000 men into the field, to so great an extent had his forces been depleted while Wellington had less than 50,000 available troops. Of the French, 3200 were killed or wounded, of the Allies 4600. On the 12th, April Soult evacuated Toulouse, six days after Napoleon the Great had snatched up a pen and scrawled his formal abdication. A moment before he had been full of fight had wanted to rally the corps of Augereau, Suchet and Soult. A year later he won back more than these. Wellington entered Toulouse on the day Soult left it and within a few hours of the

Duke of Wellington

receipt of the news from Paris of the proclamation of Louis XVIII, a monarch as incompetent as the fallen Emperor was great. History is often ironic, and Time's see-saw seldom maintains an even balance for any lengthy period.

Duke of Wellington

CHAPTER XVII
Start Waterloo Campaign (1814–15)

"I work as hard as I can in every way in order to succeed."

Wellington.

Outline of specific point of views or inside knowledge in this chapter:

79. Lady Burghersh.
80. Parliament also granted to him the sum of £400,000.
81. See "Story of Nelson," p. 195.
82. The complete Memorandum will be found in Gurwood, vol. xii., pp. 125–9.
83. "Cambridge Modern History," vol. ix. p. 619.

"I march to-morrow to follow Marshal Soult, and to prevent his army from becoming the *noyau* of a civil war in France." Thus writes Wellington to Sir John Hope on the 16th April 1814, when the white flag of the Bourbons was flying at Toulouse, and forty-eight hours after Hope had been made a prisoner during a sortie on the part of the French garrison of Bayonne.

Duke of Wellington

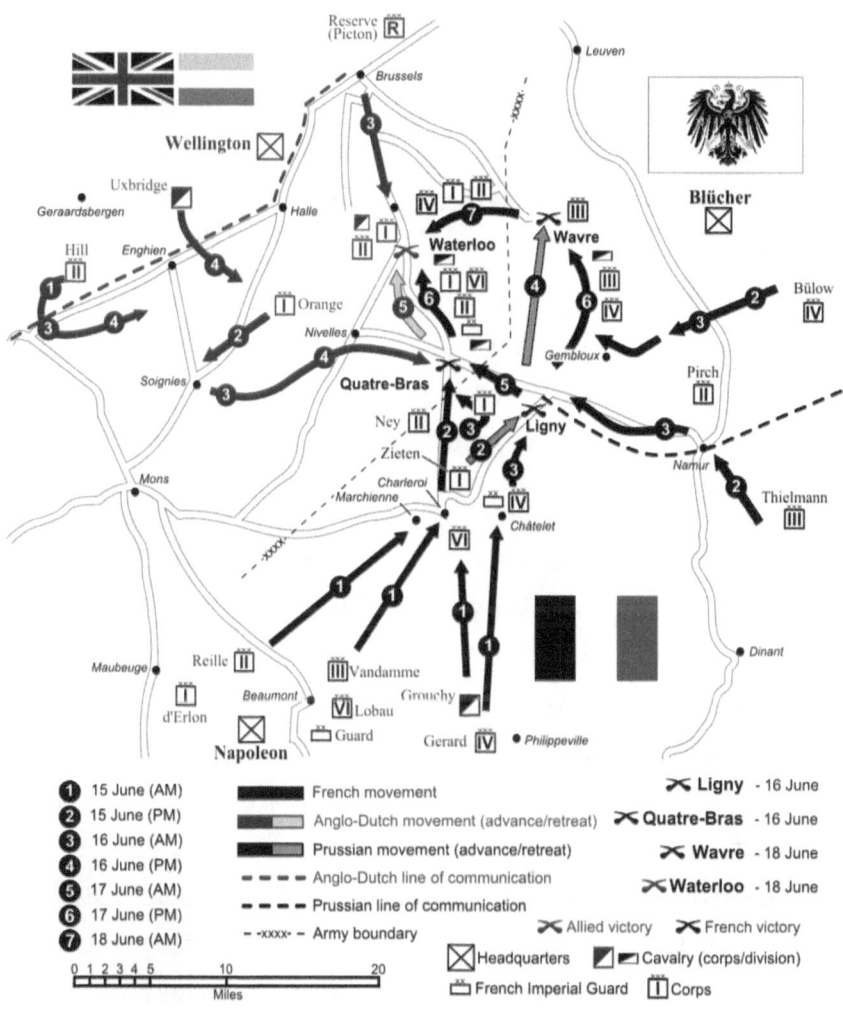

Figure 20 Map of the Waterloo Campaign

Duke of Wellington

Soult extended no right hand of welcome to Louis XVIII, and positively refused to submit to the new *régime* until he had received trustworthy information from some of Napoleon's ministers. However, he was speedily convinced of the fall of his former master, and both he and Suchet acknowledged the Provisional Government. On the 19th April, a Convention was signed by each party and Wellington for the cessation of hostilities and the evacuation of Spain. The British infantry were sent either to the homeland or on foreign service; the cavalry traversed France and crossed to England from Calais.

Wellington's work was not yet over, although his military career was closed for a time. He was appointed British Ambassador at Paris, and while he wrote to a correspondent that recent political and military events promised "to restore the blessings of peace permanently to the world," we must not suppose that he believed the abdication of Napoleon to be the herald of the millennium. When Castlereagh proposed the diplomatic post to him Wellington would have been faultlessly justified in declining it, but sufficient of his story has been told for the reader to appreciate the fact that the Hero of the Peninsula was as keenly devoted to the service of his king and the country as the Hero of Trafalgar. Whatever egotism he possessed was definitely not carried to excess. He says that he should never have thought himself qualified for the work. "I hope, however," he adds, and here the sterling qualities of the man are revealed, "that the Prince Regent, his Government, and your Lordship, are convinced that I am ready to serve him in any situation in which it may be thought that I can be of any service. Although I have been so long absent from England, I should have remained as much longer if it had been necessary; and I feel no objection to another absence in the public service if it be necessary or desirable." He says much the same thing to his brother Henry: "I must serve the public in some manner or other; and, as under current circumstances I could not well do so at home, I must do so abroad."

Those who accuse Wellington of lack of heart will do well to remember that before leaving Toulouse for Paris he wrote an appealing letter to Earl Bathurst on behalf of Sir Rowland Hill and Sir Robert Kennedy. Sir Kennedy had exerted himself to the utmost in keeping the army well supplied with provisions, and to write a letter of condolence to Hope, who was a prisoner and wounded.

But he found time to join in a few *fêtes* in honor of the Restoration, including a magnificent ball is given by Sir Charles Stewart, the British Commissioner of the Army of the Allies, where monarchs were plentiful and Society beauties abundant. "It was in the midst of this ball," the Comtesse de Boigne relates, "that the Duke of Wellington appeared for the first time in Paris. I can see him now entering the room with his two nieces, Lady Burgers [79] and Miss Pole, hanging on his arms. There were no eyes for anyone else, and at this ball, where grandeur abounded, everything gave way to military glory. That of the Duke of Wellington was brilliant and unalloyed, and a lustre was added to it by the interest that had long been felt in the cause of the Spanish nation."

He had only been in Paris six days before he set out for Madrid, viâ Toulouse, "to try whether I cannot prevail upon all parties to be more moderate, and to adopt a constitution more likely to be practicable and to contribute to the peace and happiness of the nation." He had made the proposal, and the Allies had eagerly accepted it. When he started on his journey, he was the Duke of Wellington, [80] and it was an additional cause of satisfaction to him to know that peerages had been conferred on Beresford, Hill, Cotton, Hope, and Graham, "my gallant coadjutors." He stayed in Toulouse for a couple of days, attending to details connected with the army, and again continued his journey, writing dispatches, notes of condolence, a letter requesting permission to accept the Grand Cross of the Order of St George from the Czar, and so on.

Duke of Wellington

Napoleon had released Ferdinand VII on the 13th of the previous March, and the king was now back in his capital. "I entertain a very favorable opinion of the King from what I have seen of him," Wellington writes from Madrid on the 25th May 1814, "but not of his Ministers." This opinion of Ferdinand must be taken as referring to the man and not to his methods, for he had already assumed the part of a despot to so alarming an extent that civil war was feared, hence the Duke's journey. "I have accomplished my object in coming here"; he says in the same letter, "that is, I think there will certainly be no civil war at present." But seven days later he communicates with Castlereagh in a minor key: "I have been well received by the King and his Ministers, but I fear that I have done but little good."

He left a lengthy memorandum in the hands of his Catholic Majesty, full of excellent advice, and bereft, as he said, of "all national partialities and prejudices." Commerce, the colonies, domestic interests, and finance are all touched upon in a sane, straightforward way, obviously with the intention of promoting "a good understanding and cementing the alliance with Great Britain," but valuable quite apart from any motive that might be construed as selfish. As Wellington says in the preamble, "The Spanish nation having been engaged for six years in one of the most terrible and disastrous contests by which any nation was ever afflicted. Its territory having been entirely occupied by the enemy. The country torn to pieces by internal divisions. Its ancient constitution having been destroyed, and vain attempts made to establish a new one. Its marine, its commerce, and revenue entirely annihilated. Its colonies in a state of rebellion, and nearly lost to the mother country. It becomes a question for serious consideration, what line of policy should be adopted by His Majesty upon his happy restoration to his throne and authority." Had Ferdinand taken Wellington's well-intentioned advice to heart, Spain might have risen from her ashes. The past abuses cropped up, the Inquisition was re-established in a milder form, and troops were sent across the seas to perish in a futile endeavor to recover the Transatlantic colonies of a once glorious empire.

After returning to Paris to make arrangements for the embarkation of the British cavalry at Calais, the Duke sailed for England. When he landed at Dover on the 23rd June 1814, a salute from the batteries of the Castle welcomed him home. "About five o'clock this morning," says a contemporary writer, "his majesty's sloop-of-war, the *Rosario*, arrived in the roads, and fired a salute. Shortly afterwards, the yards of the different vessels of war were manned. A salute took place throughout the squadron, and the launch of the *Nymphen* frigate was seen advancing towards the harbour, with the

Duke of Wellington. At this time the guns from the heights and from the batteries commenced their thunder upon the boat leaving the ship; and on passing the pier-heads his Lordship was greeted with three distinct rounds of cheers from those assembled; but upon his landing at the Crosswall, nothing could exceed the rapture with which his Lordship was received by at least ten thousand persons; and notwithstanding it was so early, parties continued to arrive from town and country every minute. The instant his Lordship set foot on shore, a proposition was made, and instantly adopted, to carry him to the Ship Inn: he was borne on the shoulders of our townsmen, amidst the reiterated cheers of the populace."

London went wild with excitement

When he arrived, and at Westminster Bridge, the mob took the horses from his carriage and dragged it along in triumph. On the 28th, he took his seat for the first time in the House of Lords. He must have appeared a good figure as, clad in his Field Marshal's uniform under a peer's robes, he was introduced by the Dukes of Beaufort and Richmond. Lord Chancellor expressed the sentiments of the House. He refrained from attempting to state the "eminent merits" of his military character, "to represent those brilliant actions, those illustrious achievements, which have attached immortality to the name of Wellington, and which have given to this country a degree of glory unexampled in the annals of this kingdom. In thus acting, I believe I best consult the feelings which evince your Grace's title to the character of a truly great and illustrious man"; and the Duke replied, in a short speech, attributing his success to his troops and general officers. A little later a deputation from the Lower House waited upon Wellington to offer him the congratulations of the Commons, and he attended in person to return thanks. The whole House rose as he entered. After a short speech, the Speaker made an eloquent and touching address.

"It is not ... the grandeur of military success," he said, "which has alone fixed our admiration or commanded our applause. It has been that generous and lofty spirit which inspired your troops with unbounded confidence and taught them to know that the day of battle was always a day of victory. That moral courage and enduring fortitude which, in perilous times, when gloom and doubt had beset ordinary minds, stood nevertheless unshaken; and that ascendancy of character which, uniting the energies of jealous and rival nations, enabled you to wield at will the fate and fortunes of mighty empires....

"It now remains only that we congratulate your Grace upon the high and essential mission on which you are about to proceed, and we doubt not that the same splendid talents, so conspicuous in war, will maintain, with equal authority, firmness, and temper, our national honor and interests of peace."

Wellington was made a Doctor of Laws by the University of Oxford, as Nelson had been before him, [81] he received the freedom of the City of London in a gold casket, and a magnificent sword—in a word, he was the country's Hero.

The time at his disposal was short and fully occupied. He left London on the 8th August for Paris, travelling by way of the Netherlands, where he inspected the frontier from Liège along the Meuse and the Sambre to Namur and Charleroi. Then by Mons to Tournay and the sea with a view to determining how Holland and Belgium, now united into one kingdom, could be placed in an adequate state of defence for future service should circumstances dictate. He also noted some of the most advantageous positions, including "the entrance of the *forêt* de Soignes by the high road which leads to Brussels from Binch, Charleroi, and Namur,"—in one word, Waterloo. He realized that there were more disadvantages than advantages, but "this country must be defended in the best manner that is possible," even though it "affords no features upon which reliance can be placed to establish any defensive system." [82]

Duke of Wellington

Figure 21 Battle of Waterloo

Wellington had no hours of luxurious ease in Paris. The abolition of the slave trade, on which Great Britain had at last determined, occupied much of his attention. One has only to refer to his dispatches at this period to understand the many difficulties he had to contend within this one particular. Then there were questions of compensation for private property destroyed or damaged in the late war to be considered, of American vessels of war and privateers fitted out in French ports. What was most important of all, a diagnosis of the increasing restlessness in Paris to be made. He believed that the sentiments of the people were favorable to the Bourbon king, "but the danger is not in that quarter, but among the discontented officers of the army, and others, heretofore in the civil departments of the service, now without employment."

It would be incorrect to state that Wellington was popular in Paris, for not a few prominent military men regarded the presence of the General who had played no small part in tarnishing the glory of France as a perpetual reminder of the country's misfortunes. The people even went so far as to resent his coat of arms, in which a lion or leopard was bearing a tricolored flag. This was construed as the British lion trampling on the French national flag. There was an eagle on the Duchess's arms, which was another cause of offence. "My coach was in danger of being torn to pieces," says the Duke, and he was obliged to have the innocent bird painted out.

Duke of Wellington

Figure 22 Congress of Vienna

Was now sitting, bent on undoing the work of the Revolution so far as was possible to upholding the Divine right of kings. This is not to be wondered at considering the members of the solemn conclave, which included the Czar, the Kings of Prussia, Bavaria, Denmark and Würtemberg, the Grand Duke Charles of Baden, the Elector William of Hesse, the

Hereditary Grand Duke George of Hesse-Darmstadt, the Duke of Weimar, and Prince Eugene Beauharnais (Napoleon's step-son). The President was Metternich, the Emperor of Austria's right-hand man, the first representative of France was the wily Talleyrand, of Great Britain Castlereagh. A host of plenipotentiaries came to put their fingers into the political pie, including those of Spain, Portugal, Sweden, Norway, France, Great Britain, Prussia, Russia, Switzerland, Italy, the Pope, the Netherlands, and the smaller German States.

What with talk of projected attempts on his life and the far from Pacific doings at Vienna, the Earl of Liverpool was of the opinion that it would be advisable to get Wellington out of France as soon as possible. With this idea in view, he was offered the command of the troops in North America, an offer he bitterly resented. However, Castlereagh solved the difficulty by asking the Duke to take his place in Vienna. The proposition was made by the Foreign Secretary in a letter dated the 18th December 1814. "I do not hesitate to comply with your desire," the Duke replies. "As I mean to serve the King's Government in any situation which may be thought desirable, it is a matter of indifference to me in what stage I find your proceedings."

Duke of Wellington

When Wellington reached the Austrian capital in January 1815—destined to be the greatest year in modern European history—he found that the wolves in sheep's clothing had almost concluded their deliberations. Russia, supported by Prussia, was intent upon securing Poland, a plan bitterly opposed by Great Britain and Austria. France was wishful for Holland and Belgium. The quarrelling suddenly ceased when, on the 7th March, Metternich received the most startling news. Napoleon, King of Elba, had left his little island state, landed on the French coast, and was marching in the direction of Paris! Wellington heard on the same day from another source, and immediately communicated the little news detailed to him by Lord Burghersh to the Emperor of Austria and the Czar.

"I do not entertain the smallest doubt that, if unfortunately, it should be possible for Buonaparte to hold at all against the King of France, he must fall under the cordially united efforts of the Sovereigns of Europe." Thus writes the Duke to Castlereagh, but in a dispatch of the same date, namely the 12th March, he shows that he entirely failed to appreciate the fascination still exercised by the name "Napoleon." "It is my opinion," he writes, "that Buonaparte has acted upon false or no information, and that the King will destroy him without difficulty, and in a short time." We know that the ex-Emperor's reception was at first somewhat lukewarm, but as he marched towards the capital it assumed the form of a triumphal procession, with Ney and the 6000 men who were "to bring him back in an iron cage" as enthusiastic followers. The inhabitants of the south alone refused to recognize the former Emperor of the French.

Far from Louis XVIII destroying Napoleon "without difficulty," that brave monarch left France to its own devices on the 19th March, the day before his predecessor and successor reached the Tuileries. "What did he do in the midst of the general consternation of Paris?" asks the Baron de Frénilly. "He acted. A great crowd saw him in the morning proceeding in pomp with Monsieur to the Chamber of Deputies. He was seen to enter and throw himself into his brother's arms, with the solemn promise to remain in Paris and be buried under the ruins of the monarchy; and on the following day the population learnt that he had fled in the night by the road to Flanders!" Soult, now Minister of War, apparently under the impression that an Army Order would tend to dispel any affection the soldiers might feel towards their former Head, issued the most stupid of ridiculous proclamations. "Bonaparte," it reads in part, "mistakes us so far as to believe that we are capable of abandoning a legitimate and beloved sovereign to share the fortunes of one who is nothing more than an adventurer. He believes this—the idiot! —and his last act of folly is a convincing proof that he does so."

Without loss of time the Fifth Coalition was formed, Great Britain, Russia, Austria and Prussia entering into a treaty on the 17th March. Where each of them guaranteed to put 150,000 men in the field against "the enemy and disturber of the peace of the world," Great Britain, as usual, financing the Allies, this time to the enormous extent of £5,000,000.

With commendable dispatch, Napoleon formed a new ministry and began to marshal his troops, which at first numbered 200,000 and eventually 284,000, excluding a quarter of a million of men for national defense. "It was the finest army," writes Professor Oman, "that Napoleon had commanded since Friedland, for it was purely French and was composed almost entirely of veterans; but it was too small for its purpose." [83] Murat, King of Naples, precipitated matters by invading the Papal States, and failed at the hands of Austria, thereby robbing his brother-in-law of his only possible ally. But this was finished by the beginning of May, over a month before Napoleon started for the front, leaving 10,000 of his none-too-numerous troops to quell an outburst of royalist enthusiasm in La Vendée, ever the most warlike province of France and apt to flame into insurrection on the slightest provocation.

Duke of Wellington

1. Arthur Wellesley, 1st Duke of Wellington
2. Joaquim Lobo Silveira, 7th Count of Oriola
3. António de Saldanha da Gama, Count of Porto Santo
4. Count Carl Löwenhielm
5. Jean-Louis-Paul-François, 5th Duke of Noailles
6. Klemens Wenzel, Prince von Metternich
7. André Dupin
8. Count Karl Robert Nesselrode
9. Pedro de Sousa Holstein, 1st Count of Palmela
10. Robert Stewart, Viscount Castlereagh
11. Emmerich Joseph, Duke of Dalberg
12. Baron Johann von Wessenberg
13. Prince Andrey Kirillovich Razumovsky
14. Charles Stewart, 1st Baron Stewart
15. Pedro Gómez Labrador, Marquis of Labrador
16. Richard Le Poer Trench, 2nd Earl of Clancarty
17. Wacken (Recorder)
18. Friedrich von Gentz (Congress Secretary)
19. Baron Wilhelm von Humboldt
20. William Cathcart, 1st Earl Cathcart
21. Prince Karl August von Hardenberg
22. Charles Maurice de Talleyrand-Périgord
23. Count Gustav Ernst von Stackelberg

Figure 23 Congress of Vienna Participants

CHAPTER XVIII
Ligny and Quatre Bras (1815)

"I go to measure myself with Wellington."

Napoleon.

Outline of specific point of views or inside knowledge in this chapter:

84. "The Campaign of 1815, chiefly in Flanders," by Lieut.-Colonel W. H. James, P.S.C., pp. 14–15.
85. "Cambridge Modern History," vol. ix. p. 625. See also "The Life of Napoleon I" by J. Holland Rose, Litt.D., vol. ii. p. 455.
86. James, p. 27.
87. James, p. 100.
88. Croker, vol. iii. p. 173.
89. This interesting relic still exists.
90. Rye.
91. Disbanded in 1816.

Napoleon left Paris at dawn on the 12th June and travelled to Laon. His troops were divided into the Army of the

North, intended for the invasion of Belgium, which totalled a little over 124,000; the Army of the Rhine, commanded by Rapp, about 20,000, with a reserve of 3000 National Guards. Le Courbe's corps of observation, watching the passes of the Jura, about 8000. The Army of the Alps, with Suchet, some 23,000; a detachment, under Brune, guarding the line of the Var, 6000; the 7th Corps, watching the line of the Pyrenees, 14,000, in two sections under Decaen and Clausel. The Army of the North was distributed at Lille, Valenciennes, Mézières, Thionville, and Soissons, under D'Erlon, Reille, Vandamme, Gérard and Lobau respectively; the Imperial Guard near Paris, and the Reserve Cavalry, under Grouchy, between the Aisne and the Sambre.[84] Soult was chief of the staff, an appointment not particularly happy.[85]

In Belgium

Figure 24 Charge of the Scots Greys at Waterloo

There was the nucleus of an army, consisting of some 10,000 soldiers, mostly British. Wellington arrived at Brussels on the 5th April, with the formidable task in hand of organizing a substantial body to oppose the returned Exile. He managed it, but the result was almost as motley a crowd of fighting men as Napoleon had for his disastrous Russian campaign. Wellington bluntly called them "not only the worst troops but the worst-equipped army, with the worst staff that was ever brought together." There were Hanoverians, Belgians, Dutch, Brunswickers, and Nassauers, as well as men of his own country. The 1st Corps, under the Prince of Orange, totalled 25,000, with headquarters at Braine-le-Comte. The 2nd Corps, commanded by Lord Hill, numbered 24,000, with headquarters at Ath; the Reserve Corps, with the Duke at Brussels, 21,000; the Cavalry, under the Earl of Uxbridge, 14,000; in the garrisons were 12,000, and the artillery and engineers reached 10,000—grand total 106,000. 86 The Prussian Army, commanded by Blücher, reached 124,000 men, some few thousands of whom were already in Belgium in March. It was made up of four corps stationed at Charleroi, Namur, Ciney, and Liége, with headquarters at Namur. Both armies were in touch with each other, although distributed over a large extent of territory. It was intended that 750,000 men should be available for the invasion of France, but none of the other allies was ready. Napoleon acted promptly, his idea being to deal with each separately and drive them back on their bases before they were able to concentrate. He would then turn on the Austrians before the Russians were ready.

Napoleon succeeded in concentrating the Army of the North without prescribed particulars of his movements reaching either Wellington or Blücher. On the 15th June, he was across the frontier and had made a preliminary success by driving Ziethen, who commanded Blücher's first corps, from the banks of the Sambre, gaining the bridges, and securing Charleroi.

Figure 25 Jan Willem Pieneman: The Battle of Waterloo (1824). Duke of Wellington, centre, flanked on his left by Lord Uxbridge in hussar uniform. On the image's far left, Cpl. Styles of the Royal Dragoons flourishes the eagle of the 105eme Ligne.

The Emperor followed the Prussians to within a short distance of Gilly, where the French right wing defeated them with the loss of nearly 2000 men. The enemy then fell back in the direction of Ligny, and Napoleon made his headquarters at Charleroi. Meanwhile Ney, who had only arrived in the afternoon, was given charge of Reille's and D'Erlon's corps, and it is usually contended that he had told Lefebvre-Desnoëttes to reconnoiter towards Quatre Bras, then held by some 4500

Nassau troops, commanded by Prince Bernard of Saxe-Weimar.

Lefebvre first encountered the enemy at the village of Frasnes, some twenty-three miles from Brussels and covering Quatre Bras, where about 1500 men were stationed, who fell back towards Quatre Bras. The French General occupied the village in the evening after an indecisive action.

When information reached Wellington from Ziethen, vague because it was dispatched early in the morning, he ordered the majority of the troops at his disposal to be "ready to move on the shortest notice," and a few only were told to change the positions they then occupied. He issued his final instructions at 10 p.m., and then went to the ball given by the Duchess of Richmond, who had invited some of the non-commissioned officers and privates in order "to show her Bruxelles friends the *original Highland dance*," as Wellington afterwards averred. The Commander-in-Chief was quite easy in his mind, for he had done all that it was possible for him to do, and his appearance at such a festivity tended to allay the anxiety of the inhabitants as to Napoleon's movements. Surely the capital was safe if Wellington was so unconcerned as to go to a dance?

Duke of Wellington

Figure 26 A British square puts up dogged resistance against attacking French cavalry

There was a sound of Revelry by night. Belgium's Capital had gathered then Her Beauty and her Chivalry. The bright lamps shone o'er fair women, and brave men. A thousand hearts beat happily; and when Music arose with its voluptuous swell, Soft eyes looked love to eyes which spake again; And all went merry as a marriage-bell: But Hush! Hark! A deep sound strikes like a rising knell!

Duke of Wellington

Did ye not hear it? No: 'twas but the wind; Or the car rattling o'er the stony street: On with the dance! Let Joy be unconfined: No sleep till morn when Youth and Pleasure meet to chase the glowing Hours with flying feet: But, Hark! That heavy sound breaks in once more; As if the clouds its echo would repeat; And nearer, clearer, deadlier than before! Arm! Arm! It is; it is; the Cannon's opening Roar!

Had Byron pictured the entry of Lieutenant Webster with a dispatch for the Prince of Orange he would have been more literally, if not artistically, correct. The nineteen words which the Prince read were momentous: "Captain Baron von Gagern has just arrived from Nivelles, reporting that the enemy had pushed up to Quatre Bras." It had been written at Braine-le-Comte at 10.30 p.m. No further instructions were issued by Wellington, for he had already arranged "for a movement tending to draw the right of his forces in towards the left." [87] He, therefore, remained at the ball until about 2 a.m., on the 16th, and then went to bed.

He left the city very early and arrived at Quatre Bras about 10 o'clock. There he found, according to his own statement, [88] "the Prince of Orange with a small body of Belgian troops, two or three battalions of infantry, a squadron of Belgian dragoons, and two or three pieces of cannon which had been at the Quatre Bras—the four roads—since the preceding evening. It appeared that the picket of this detachment had been touched by a French patrol. There was some firing, but tiny; and of so little importance that, after seeing what was doing, I went on to the Prussian army, which I saw on the ground, was assembling upon the field of St Amand and Ligny, about eight miles distant.

"I reached the Prussian army; was at their headquarters; stayed there a considerable time; saw the army formed; the commencement of the battle; and returned to join my own army assembled and assembling at the Quatre Bras. I arrived then at Quatre Bras a second time on that day, as well as I recollect, at about two or three o'clock in the afternoon.

"The struggling fire there had continued from morning; the Prince of Orange was with the line troops still in the same position. I was informed that the army was collecting in a wood in front. I rode forward and reconnoitered or examined their position according to my usual practice. I saw clearly a vast body of men assembled, and a Maréchal reviewing them, according to their customary practice, preparatory to an attack. I heard the usual cries: 'En avant distinctly*!* en avant! L'Empereur récompensera celui qui s'avancera!'

"Before I quitted the Prince of Orange, some of the officers standing about had doubted whether we should be attacked at this point. I sent to the Prince of Orange from the ground on which I was standing, to tell him that he might rely upon it that we should be attacked in five minutes and that he had better order the retreat towards the main position of the light troops and guns which were in front. These could make no resistance to the fierce attack about to be made upon us. These were accordingly withdrawn, and in less than five minutes we were attacked by the whole French army under Maréchal Ney. There was, in fact, no delay nor cessation from attack from that time till night. The reserves of the British army from Brussels had arrived at the Quatre Bras at this time; and the corps of Brunswick troops from the headquarters, and a division of Belgian troops from Nivelles, Braine, &c."

Ney did not begin his attack until 2 p.m. He had only Reille's corps at his disposal, and he was uncertain as to the movements of the Prussians. But when a move was made against the farm of Gémioncourt, 89 the key of the position, the 7000 troops of the Prince of Orange were speedily driven back, and the wood of Bossu also fell into the enemy's hands. It was at this precise moment that Wellington and reinforcements arrived. Picton with his strong 5th Division, although exhausted by a long march on a hot day, were ordered to retake the wood. It was then that Ney hurled his cavalry at them in a dangerous endeavor to save the situation at all costs.

"One regiment," we are told, "after sustaining a furious cannonade, was suddenly, and on three different sides, assailed by cavalry. Two faces of the square were charged by the Lancers while the cuirassiers galloped down upon another. It was a trying moment. There was a death-like silence, and one voice alone, calm and clear, was heard. It was their Colonel's, who called upon them to be 'steady.' On came the enemy—the earth shook beneath the horses' feet; while on every side of the devoted band, the corn, bending beneath the rush of cavalry, disclosed their numerous assailants. The lance-blades approached the bayonets of the kneeling front rank—the cuirassiers were within forty paces—yet not a trigger was drawn. But when the word 'Fire!' thundered from the colonel's lips, each piece poured out its deadly volley, and in a moment the first files of the French lay before the square as if hurled by a thunderbolt to the earth. The assailants, broken and dispersed, galloped off for shelter to the tall rye; while a stream of musketry from the British square carried death into the retreating squadrons." At length, Maitland's division of the Guards secured possession of the wood. The French continued to attack with ardor, but with the close of day Ney fell back upon the road to Frasnes. The British remained at Quatre Bras.

Many brave incidents are recorded. A soldier of the 92nd was wounded in the thigh. After having been attended by a surgeon the medical man dismissed him by saying, "There is now no fear of you, get slowly behind." His reply, unpremeditated and instant, was, "The presence of every man is necessary," and calmly went back to his post, from which he never returned. Another fellow had his knapsack carried away from his shoulders by a cannon-ball. He caught it by a quick movement before it reached the ground. Wellington happened to be near, and the incident afforded him considerable amusement.

The 2nd Battalion of the 69th Regiment of Foot [91] had its flag captured, the particular symbol being wrenched from the hands of the officer almost at the same time as the attempted capture of another color during a charge of French cuirassiers. The latter was preserved, although the bearer received severe wounds. Ensign Christie also saved a flag by throwing himself upon it. As the soldier lay on the ground a portion of the color was torn off by the lancer who had made the attempt, but before the prize could be taken the captor was shot and the piece recovered.

Duke of Wellington

Wellington lost 4600 killed and wounded at Quatre Bras, but although he remained in possession of the field, Ney's grim determination had precluded him from sending reinforcements to Blücher, who had been contesting Napoleon at Ligny, eight miles distant. On the other hand, the Emperor had ordered his Marshal to assist him at Ligny after having dispersed the troops at Quatre Bras, which he was far too heavily engaged all day to do, although it is conceivable that had he began his attack earlier he might have carried out his Chief's instructions. Napoleon himself delayed his attack because some of his soldiers had not arrived, and when he was ready to open fire, he totaled 20,000 men less than the Prussian commander. D'Erlon's troops, forming Ney's reserve, were sent for, but before they were pressed into service, were ordered back by Ney to assist him. Had D'Erlon remained, Napoleon would have won a decisive victory. As it was, Blücher's losses reached over 20,000, and he narrowly averted capture. He was thrown during a cavalry charge and seriously injured. The Prussians abandoned the field, retiring towards Wavre, which enabled them to be in touch with Wellington, and where they later formed a junction with the fresh corps of Bülow. Napoleon, who had 11,000 men killed or wounded, was convinced that the enemy would fall back on Namur and Liége, which would enable him to deal with Wellington alone on the morrow.

Duke of Wellington

Napoleon was unusually lethargic on the following day and squandered much precious time until noon in various stupid ways. He then ordered Grouchy, with 33,000 men, to ascertain the position occupied by the Prussians, while he joined Ney. Wellington, hearing of Blücher's defeat and realizing that every moment was of value, had evacuated Quatre Bras almost without interference, although Napoleon's cavalry came up with the British rear and the situation was saved by a propitious downpour of rain which soon transformed road and field alike into a quagmire. Still, Napoleon chased in pursuit to the ridge of Waterloo, where a heavy cannonade put an end to his advance. The Emperor made his headquarters at the Caillou farm, retiring to rest only after he had seen that Wellington's army was not retreating from the position it had taken up at Mont St Jean. Unable to sleep, he again satisfied himself that the army he was bent on shattering had not stolen a march on him, and when the 18th June dawned he was still gazing in the direction of Wellington and his men.

CHAPTER XIX
Waterloo (1815)

"The history of a battle is not unlike the history of a ball. Some individuals may recollect all the little events of which the great result is the battle lost or won; but no individual can recollect the order in which, or the exact moment at which they occurred, which makes all the difference as to their value or importance."

Wellington.

Outline of specific point of views or inside knowledge in this chapter:

92. Rose's "Napoleon," vol. ii. p. 487–8.
93. Rose's "Napoleon," vol. ii. p. 488.
94. Comte Charles van der Burch is the present owner of Hougoumont.
95. Now the Grenadier Guards.
96. Rose, vol. ii. p. 496.
97. "Some of this brigade, particularly the 5th Military, had behaved with great gallantry on the 16th, at Quatre Bras."—Cotton's, "A Voice from Waterloo," p. 56.
98. General Gascoigne in the House of Commons, the 29th June 1815.

99. "Someone asked whether the French Cuirassiers had not come up very well at Waterloo? 'Yes,' he (Wellington) said, 'and they went down very well too.'"—Croker, vol. i. p. 330.
100. *I.e.* the guns were not removed, the artillerymen working them until the last moment and then seeking refuge in the nearest square, to resume their former position when the enemy began to retire.
101. "Cambridge Modern History," vol. ix. p. 639.

The British General had already sent word to Blücher that he was prepared to fight Napoleon if the Prussian Commander could see his way to send him one army corps, the assistance of which he deemed imperative. He did not receive a reply until the early hours of the 18th, and it was in the affirmative. He promised Bülow's corps, which would march at daybreak against Napoleon's right, followed by that of Pirch. Those of Thielmann and Ziethen would also be sent provided the presence of Grouchy did not prevent. As events turned out, Thielmann's corps was sufficient for this purpose of defending the Dyle against the French Marshal. The first report from Grouchy received by Napoleon on the 18th was sent from Gembloux. It contained news up to 10 o'clock on the previous night, namely, that part of the Prussians had retired towards Wavre, that Blücher with their centre had fallen

Duke of Wellington

back on Perwez, and a column with artillery had moved in the direction of Namur. 92

From another source Napoleon gathered that three bodies of the Prussians were concentrating at Wavre, vital information which, for some inexplicable reason, he did not dispatch to Grouchy until 10 a.m. We have it on the authority of Foy, who had fought in Spain, that at daybreak Napoleon was told that the junction of Wellington and Blücher was possible. This he refused to believe, stating that such an event could not take place before the 20th. Grouchy was ordered to keep up his communications with Napoleon, pushing before him "those bodies of Prussians which have taken this direction and which may have stopped at Wavre." Also, he was to "follow those columns which have gone to your right." Dr Holland Rose pronounces these instructions as far from clear: "Grouchy was not bidden to throw all his efforts on the side of Wavre, and he was not told whether he must attack the enemy in that town, or interpose a wedge between them and Wellington, or support Napoleon's right. Now Napoleon would certainly have prescribed a critical concentration of Grouchy's force towards the north-west for one of the last two objects, had he believed Blücher about to attempt a flank march against the superior French army. Apparently it had not yet entered his thoughts that so daring a step would be taken by a foe whom he pictured as scattered and demoralized by defeat." **93**

As Houssaye, one of the most painstaking of historians, says, "It is necessary to walk over the ground, to perceive it's relentlessly undulating formation, similar to the billows of a swelling sea." I have followed his advice for the purposes of the present volume. The configuration of the land near the ridge, which was at first the right center of the Allies, has been altered somewhat by the excavation of earth for the celebrated but unsightly Lion Mound. Otherwise, the field is much the same as it was in 1815. Major Cotton, who fought at Waterloo, tells us that on the day of the battle there were "splendid crops of rye, wheat, barley, oats, beans, peas, potatoes, tares, and clover; some of these were of great height. There were a few patches of ploughed ground." Exactly the same is true now. The soil which covers the moldering bones of many a hero yields a mighty harvest. Practically every inch of it is either pasture or under cultivation.

By far the most interesting point is the château and farmhouse of Hougoumont. The buildings were then over two centuries old, and were erected to (or "intending to") defense, the garden being strongly walled on the south and east sides. In 1815, it was replete with orchard, outbuildings, banked-up hedges, and a chapel. Additional loop-holes were made by

Duke of Wellington

Wellington's orders, a scaffold was erected so that the troops could discharge their muskets over the wall, and the flooring over the south gateway hastily taken up "to enable our men to fire down upon the enemy should they force the gate which had been blocked up." This strong strategic position on the right flank of the Allied line was admirably defended by the light companies of the second battalions of Coldstream and Foot Guards.

The battle began with a protracted but successful attack by Napoleon's brother, Jerome, on the Nassauer and Hanoverian troops, which held the wood in front of the château. Proud of his triumph, and contrary to the Emperor's orders, three attempts to assault the grim building were made and failed. At length Napoleon made up his mind to bombard Hougoumont, which he did to some effect, the chapel and barn blazing fiercely, but its brave defenders never surrendered. Four days after the battle, Major W. E. Frye visited the spot. "At Hougoumont," he says, "where there is an orchard, every tree is pierced with bullets. The barns are all burned down, and in the courtyard it is said they have been obliged to burn upwards of a thousand carcasses, an awful holocaust to the War-Demon."

On one of the outside walls of the little chapel, within which many gallant soldiers breathed they're last, is a bronze tablet erected in 1907, "to the memory of the brave dead by His Britannic Majesty's Brigade of Guards and by Comte Charles van der Burch." 94 The sacred building has been thoroughly repaired, and the interior white-washed. The latter is to be regretted but was incumbent owing to the vandalism of visitors whose one aim in life apparently is to carve or scrawl their names upon monuments and buildings.

Within easy distance is the farm of La Haye Sainte, still used for the purpose for which it was originally built. If the thick wall which abuts the road to the Belgian capital is not so heavy-duty as it was—and there are signs of recent repair—the most cursory examination is sufficient to prove that it offered a tough resistance. La Haye Sainte was the key of the allied position. It was Napoleon's ambition to secure it, together with the farm of Mont St Jean, so that the enemy's communications with Brussels and Blücher might be cut off. Wellington had unwisely left the defense of La Haye Sainte to a mere handful of men, 376 in all, of the King's German Legion. The buildings were attacked on all sides during the eventful day, and it was not until 6.30 p.m., or thereabouts, when ammunition was exhausted, that the place fell. Professor Oman points out that the Emperor should have sent the Guard "to the front *en masse*" the moment that happened.

Duke of Wellington

This he did not do, and a golden opportunity was irretrievably lost.

A little farther up the road, and on the opposite side, is a long, white-washed building known as the farm of Mont St Jean, the hamlet of that name being the center of the position. This was the primary hospital of the Allies, and once a monastery of the Knights Templar. Near here the Iron Duke drew up the second line of his reserve, with three reserve batteries. Napoleon's headquarters were on the Belle Alliance height, where he spent the greater part of the day, and where he kept the Imperial Guard, nicknamed "The Immortals," in reserve until after 7 p.m. It was not until evening had given place to the night that the French troops were routed.

When you enter La Belle Alliance Inn, you are shown "the historical chamber of Napoleon." But let me warn visitors who admire the grotesque painting of "Wellington Meeting Blücher," which is nailed to the outside wall, against believing that this is the scene of the memorable incident. Mutual congratulations had taken place before La Belle Alliance was reached, the Duke turning off the high road leading to the village of Waterloo to meet the rugged old soldier when he described him surrounded by his staff.

Within a few hundred yards of the inn is a memorial "to the last combatants of the Grand Army." It is by far the most artistic and opposite of the several monuments on the field. With shattered wings, the French eagle guards the tattered standard of "The Immortals" on the ground where the last square of Napoleon's legendary Guard was cut down. "It was a fatality," said the fallen Emperor of the West, "for, in spite of all, I should have won that battle."

Having glanced at the chief objects of interest in the field of Waterloo, we must now turn our attention to the dominant features of the fight. It is not proposed to enter into minute details, or to discuss the many vexed points which have been raised from time to time. Sir Harry Smith, who took part in the battle, issued a word of warning years ago, which we shall endeavor to bear in mind. "Every moment was a crisis," he said, "and the controversialists had better have left the discussion on the battle-field."

The morning of Sunday, the 18th June 1815, was dismal, foggy, and wet. No "Sun of Austerlitz" pierced the low clouds that scudded across the sky. The state of the weather being then, as now, a universal topic of conversation and record, we have abundant documentary evidence on the subject. After this uninteresting, but not unimportant, fact, our witnesses, to a large extent, break down. It would be much easier to detail what we do not know about the battle of Waterloo than to put down in black and white what may be regarded

Duke of Wellington

as indisputable truth. For instance, there is a striking disparity between the times at which eye-witnesses assert the first cannon was fired. The Duke says eleven o'clock, Napoleon two hours later. Alava mentions half-past eleven, Ney one o'clock. Lord Hill, who used a stop-watch, avers with some semblance of authority, 11.50 a.m. The Emperor undoubtedly delayed giving battle. Soult issued an order between 4 and 5 a.m. for the army to be "ready to attack at 9 a.m." At 11 o'clock Napoleon stated that the soldiers were to be "in battle array, about an hour after noon." Scores of volumes and thousands of pages have been devoted to the subject of the conflict which concluded Napoleon's military career. It seems fairly reasonable to suppose that after the lapse of one hundred years any further research, although it may throw valuable sidelights on the battle, will not solve the problem. We have the Duke's assurance that "it is impossible to say when each essential occurrence took place, nor in what order." To a would-be narrator he wrote, "I recommend you to leave the battle of Waterloo as it is." He wrote to another, "the Duke entertains no hopes of ever seeing an account of all its details which shall be true," and again, "I am really disgusted with and ashamed of all that I have seen of the battle of Waterloo." That being so, it is not purposed to view it from a critical standpoint, but merely to detail the leading events of the day as these are generally accepted.

Duke of Wellington

The ridge on which Napoleon posted his forces in three lines, with a reserve of 11,000 troops behind the center, was opposite Mont St Jean, with the farm of La Belle Alliance, on the high road from Charleroi to Brussels, in the center. The number of troops at his immediate disposal was 74,000, and they occupied a front of about two miles. The spectacle as the French troops deployed "was magnificent". Napoleon afterwards averred, not without a touch of imagination, "and the enemy, who was so placed as to behold it down to the last man, must have been struck by it: the army must have seemed to him double in number what it really was."

Just before the conflict began, there was scarcely a mile between the French and Wellington's 67,000 troops, who were posted on the high roads leading respectively from Charleroi and Nivelles to Brussels, just in front of Mont St Jean, where they cross. His right extended to Merbe Braine, and his left, which he considered was protected by the advancing Prussians, was to the westward of the high road. The army was drawn up in two lines, the second of which was composed entirely of cavalry under Lord Uxbridge. To the left Wellington held the villages of La Haye, Papelotte, and Smohain; Hougoumont was before his right, La Haye Sainte in front of the center, and all were occupied by his troops. His reserves, numbering about 17,000, were unseen by the enemy owing to the formation of the ground, and were posted to his center and

Duke of Wellington

right. At the rear was the forest of Soignes, through which he could retreat if necessary. At Hal, nine miles away, the corps of Prince Frederick of Orange and two brigades of Colville's division, some 14,000 men in all, were stationed. They took no part in the battle. Their presence at so great a distance was due to the somewhat unnecessary anxiety of Wellington, as after events proved, regarding the right flank. The Allies' guns numbered 156, those of the French 246.

Wellington displayed unusual activity and commanded in person. Napoleon relied on his subordinates far more than was usual with him and seemed to take a little direct interest in the fighting. He sat in226 a chair; his antagonist rode about throughout the day. "It is hardly possible," says Lord William Lennox, one of his aides-de-camp, "to describe the calm manner in which the hero gave his orders and watched the movements and attacks of the enemy. In the midst of danger, bullets whistling close about him, candid shot ploughed the ground he occupied, and men and horses falling on every side, he sat upon his favorite charger, Copenhagen, as collectedly as if he had been reviewing the Household Troops in Hyde Park."

Although it would be foolish to endeavour to accurately divide the battle into different parts as one would do a sum. This is for the simple reason that fighting was going on all the time and the entire armies of the combatants were not used at any given moment. There were five more or less distinct attacks made by the French which may be useful as keys, viz.: (1) the diversion against Hougoumont, which opened the battle; (2) the attack against Wellington's left-centre; (3) the cavalry attack on the right centre; (4) a second attempt with infantry and cavalry having a similar object, and an infantry attack against the enemy's left; (5) the charge of the Imperial Guard.

The main attack was on the British left and centre. However, by way of diversion an attempt was first made on the wood of Hougoumont, which was carried after so determined a resistance that Alison. The historian afterwards counted no fewer than twenty-two shots in a tree less than six inches in diameter; fairly conclusive evidence of the death-dealing shower faced by the Nassauers and Hanoverians defending the copse and of the vigour of the enemy. The château was then attacked, contrary to orders, by Jerome Bonaparte, and brilliantly repulsed by the light companies of the second battalions of Coldstream and Foot Guards. Several attempts were afterwards made to secure the place, and dead bodies lay piled in heaps, but those within held it from the beginning to the end of the battle, although Wellington found it necessary to reinforce the men who were

Duke of Wellington

upholding Britain's honor so determinedly. The orchard was captured and regained, howitzers battered the big walls, the building was set on fire, the door of the courtyard was burst open and shut in the face of the French. These deeds were performed by the fellows of whom Napoleon had spoken earlier in the day as "breakfast for us!"

It was not until about half-past one o'clock, when a black, moving mass was seen on the wooded heights of St Lambert that the Emperor certainly bestirred himself. Napoleon looked through his glass in the direction of the object on which nearly all eyes were strained. Some of his officers thought it a body of troops, some suggested Prussians, others Grouchy. "I think," remarked Soult, "it is five or six thousand men, probably part of Grouchy's army." In reality, it was the advanced guard of Bülow's troops, and the Emperor shortly afterwards heard from the lips of a prisoner that at least 30,000 men were approaching to assist Wellington. However, some light horsemen were sent towards Frischermont to observe the Prussians, and a postscript was added to a dispatch already penned to Grouchy, begging him to "lose not an instant in drawing near and joining us, to crush Bülow, whom you will catch in the very act."

Without delay, Napoleon launched D'Erlon's corps in four columns totaling nearly 20,000 men against the enemy's left-center, Ney in command. They were supported by far too few cavalries. The Emperor's idea was to take La Haye Sainte, break through the Allied line, and gain Mont St Jean. This operation, if successful, would compel Wellington to abandon his communications with the Belgian capital and change his formation. Also, it would place the French between his army and the Prussians. Bylandt's Dutch-Belgians, who were nearest to the enemy and consequently more exposed to the covering fire of seventy-eight guns and the shots of the skirmishers, took to their heels as D'Erlon's divisions, frantically yelling "*Vive l'Empereur!*" approached the front line. 97 The brigades of Pack and Kempt were at once brought forward by Picton. They stood firm and poured death into the oncoming columns, receiving them, as they appeared on the crest of the ridge, with fixed bayonets.

Duke of Wellington

The Earl of Uxbridge

Figure 27 Earl of Uxbridge

Duke of Wellington

Who was in command of the cavalry, realizing that the position was still one of considerable danger, then ordered Ponsonby's Union Brigade—the 1st Royal Dragoons, Scots Greys, and Inniskillings—to charge. It burst upon the French with tremendous force and decided the issue.

Several thousands of the enemy were killed or wounded, 8000 taken prisoners, some guns silenced, and two eagles captured.

The story of how the colors of the French 45th Regiment were secured by Serjeant Ewart, a magnificent specimen of a man, over six feet in height, who served in the Greys, is best told in his own proper language. "It was in the charge I took the eagle from the enemy," he says; "he and I had a hard contest for it; he made a thrust at my groin, I parried it off and cut him down through the head. After this a lancer came at me; I threw the lance off by my right side, and cut him through the chin and upwards through the teeth. Next, a foot-soldier fired at me, and then charged me with his bayonet, which I also had the good luck to parry, and then I cut him down through the head; thus ended the contest. As I was about to follow my regiment, the General said, 'My brave fellow, take that to the rear; you have done enough till you get quit of it.' I took the eagle to the ridge, and afterwards to Brussels."

Duke of Wellington

We also have the record of Captain Clark Kennedy, of the Royal Dragoons, regarding the capture of the eagle of the 105th Regiment. "I was," he relates, "in command of the centre squadron of the Royal Dragoons in this charge. While following up the attack, I perceived a little to my left, in the midst of a body of infantry, an eagle and colour, which the bearer was making off with towards the rear. I immediately gave the order to my squadron, 'Right shoulders forward!' At the same time leading direct upon the eagle and calling out to the men with me to secure the colour. The instant I got within reach of the officer who carried the eagle, I ran my sword into his right side, and he staggered and fell. He did not reach the ground on account of the pressure of his companions: as the officer was in the act of falling, I called out a second time to some men close behind me, 'Secure the colour, it belongs to me.' The standard covered, Corporal Styles, and several other men rushed up, and the eagle fell across my horse's head against that of Corporal Styles's. As it was falling, I caught the fringe of the flag with my left hand, but could not at first pull up the Eagle: at the second attempt. However, I succeeded. Being in the midst of French troops, I struggled to separate the eagle from the staff, to put it into the breast of my coatee, but it was too firmly fixed. Corporal Styles said, 'Sir, don't break it'; to which I replied, 'Very well; carry it off to the rear as fast as you can.' He did so."

Unfortunately, the victors were too eager, and instead of returning they continued until they were in the French lines, thus enabling Napoleon to "turn the tables." His reserve squadrons robbed the British ranks of 1000 brave men, including Ponsonby. More would have fallen had not Vandeleur's Light Cavalry Brigade checked the enemy.

Picton was dead, shot at the head of his men. "When you hear of my death," the latter had previously remarked to a comrade, "you will hear of a damn good day." He had "fallen gloriously at the head of his division, maintaining a position which, if it had not been kept, would have altered the fate of the day." 98

The Highlanders

The following account of the magnificent charge of Ponsonby's Union Brigade, from the pen of James Armour, Rough-rider to the Scots Greys, who took part in it, gives some idea of the work performed:

"Orders were now given that we were to prepare to charge. We gave our countrymen in front of us three hearty huzzas, and waving our swords aloft in the air, several swords were struck with balls while so doing; and I must not forget the Piper—

Duke of Wellington

The piper loud and louder blew, the balls of all denominations quick and quicker flew.

We're then ordered to wheel back—I think by sections, but I am not certain: infantry words of command differ from the cavalry. When they had, and were wheeling back imperfectly, we rushed through them; at the same time, they huzzaed us, calling out, '*Now, my boys—Scotland forever!*' I must own it had a stimulating effect on me. I am convinced numbers of them were knocked over by the horses: in our anxiety we could not help it. Some said 'I didna think ye wad hae sair'd me sae'—catching hold of our legs and stirrups, as we passed, to support themselves. When we got clear through the Highlanders (92nd) we were now on the charge, and a short one it was. A cross road being in our way, we leaped the first hedge gallantly; crossed the road, and had to leap over another hedge. At this time, the smoke from the firing on both sides made it so that we could not see distinctly. We had not charged far—not many yards, till we came to a column. We were pretty well together as yet, although an excessive number fell about that cross road. We were in the column in a very short time (making pretty perfect work). We still pushed forward, at least as many as could—a number had dropped off by this time—and soon came to another column. They cried out, 'Prisoners!' and threw down their arms, and stripped

Duke of Wellington

themselves of their belts (I think it is part of the French discipline to do so), and ran to our rear. Ay, they ran like hares. We still pushed on, and came upon another column; and some of them went down on their knees, calling out 'Quarter!' in a very respectful way....

"We now got amongst the guns, the dangerous guns, which had annoyed us so much. *Such slaughtering!* —men cut down and run through, horses houghed, harness cut and all rendered useless. Some, who were judges of such work, reckoned we did a magnificent job of it. Amongst the guns—I think six or seven in number, all brass—that I was engaged with, mostly all the men were cut down, and the horses, most of them, if not all, were houghed. While we were at work amongst these guns, never thinking but, when we were done with it, we would have nothing to do but to return from where we came. I must own I was very much surprised, when we began to retrace our steps, when what should we behold coming away across betwixt us and our own army but a great number of these cuirassiers and lancers, the first I ever beheld in my life, who were forming up in order to cut off our retreat; but, nothing daunted, we faced them manfully. We had none to command us now, but every man did what he could. 'Conquer or die!' was the word. When the regiment returned from the charge mentioned, the troops that I belonged to did not muster above one or two sound men (unwounded) belonging to the front rank. Indeed, the whole troop did not muster above a dozen; there were upwards of twenty of the front rank killed, and the others wounded."

Meanwhile the farm of La Haye Sainte, held by a detachment of the German Legion under Major Baring, had been the object of assault by a French division detached for that purpose, and the defenders were driven from the orchard. Reinforcements were sent forward by Wellington, followed by additional cavalry by Napoleon. The cuirassiers were met by Lord Edward Somerset's Brigade, led by the Earl of Uxbridge, and driven back after a furious struggle for supremacy. Not until the last pinch of powder was spent and several serious attempts had been made to secure the place, did the brave fellows vacate their position to D'Erlon's infantry, between six and seven o'clock.

Towards the close of the afternoon Milhaud's cuirassiers, supported by light cavalry and dragoons, attacked the British right-center. Wellington at once formed his troops in squares. When the crash came, they stood firm and unbroken.

The Commander-in-Chief lost not a moment. The heavy brigade of Lord Edward Somerset was flung against the Frenchmen and drove them from the ridge. **99** The artillery blazed destruction. Charge followed charge, thirteen in all, but the British squares remained steady, although continually reduced. Nearly all Wellington's cavalry was pressed into service, and except a separate Dutch-Belgian division, all his infantry reserve was employed. An officer who served in Halkett's brigade has left a vivid account of this part of the battle.

"Hougoumont and its wood," he writes, "sent up a large flame through the dark masses of smoke that overhung the field; beneath this cloud the French were indistinctly visible. Here a waving mass of long red feathers could be seen; there, gleams as from a sheet of steel showed that the cuirassiers were moving; 400 cannons were belching forth fire and death on every side; the roaring and shouting were indistinguishably commixed—together they gave me an idea of a laboring volcano. Bodies of infantry and cavalry were pouring down on us, and it was time to leave contemplation, so I moved towards our columns, which were standing up in square.... As I entered the rear face of our square I had to step over a body, and looking down, recognized Harry Beere, an officer of our Grenadiers, who about an hour before shook hands with me, laughing, as I left the columns.... The tear was not dry on my cheek when poor Harry was no longer thought of. In a few minutes after, the enemy's cavalry galloped up and crowned the crest of our position. Our guns were abandoned, **100** and

Duke of Wellington

they formed between the two brigades, about a hundred paces in our front. Their first charge was magnificent. As soon as they quickened their trot into a gallop, the cuirassiers bent their heads so that the peaks of their helmets looked like visors, and they seemed cased in armor from the plume to the saddle. Not a shot was fired till they were within thirty yards when the word was given, and our men fired away at them. The effect was magical. Through the smoke, we could see helmets falling, Cavaliers starting from their seats with convulsive springs as they received our balls, horses plunging and rearing in the agonies of fright and pain. As the crowds of the soldiers dismounted, part of the squadron in retreat, but the most daring remainder backing their horses to force them on our bayonets. Our fire soon disposed of these gentlemen. The main body re-formed in our front, and rapidly and gallantly repeated their attacks. In fact, from this time (about four o'clock) till near six, we had a constant repetition of these brave but unavailing charges. There was no difficulty in repulsing them, but our ammunition decreased alarmingly. At length, an artillery wagon galloped up, emptied two or three casks of cartridges into the square, and we were all comfortable....

"Though we repetitively thrashed our steel-clad opponents, we found more difficult customers in the round shot and grape, which all this time played on us with terrible effect, and fully avenged the cuirassiers. Often as the volleys created openings in our square would the cavalry dash on, but they were uniformly unsuccessful. A regiment on our right seemed sadly disconcerted, and at one moment was in considerable confusion. Halkett rode out to them, and seizing their color, waved it over his head, and restored them to something like order, though not before his horse was shot under him. At the height of their unsteadiness we got the order to 'right face' to move to their assistance. Some of the men mistook it for 'right about face,' and confronted accordingly, when old Major M'Laine, 73rd, called out, 'No, my boys, its "right face"; you'll never hear the right about as long as a French bayonet is in front of you!'"

At Planchenoit

In the rear of the French center, Lobau and his 10,000 men were doing their utmost to prevent Bülow with three times that number of troops from succoring Wellington. The French held the village for two hours, and then saw it fall into the hands of the enemy, but only for a time. Some of the Young Guard and three batteries of artillery had been sent by the Emperor to support Lobau, and when they arrived the scales again turned in favour of the French. While this body was held in check, Napoleon's infantry had suffered near

Duke of Wellington

Hougoumont, but La Haye Sainte had fallen, as already noticed. Had Napoleon sent reinforcements to Ney he might have won the battle, but he hesitated to use his remaining reserves. Before anything further was done, Wellington had made different dispositions, and Ziethen's men, who had marched by way of Ohain, were on the field.

Six battalions of the Middle Guard and two battalions of the Old Guard were at last sent forward. [101] As they crossed the open ground between Hougoumont and the high road the artillery played sad havoc with some of them, but behind the crest of the ridge was Maitland's brigade of the Foot Guards, led by Wellington, himself. "Up, Guards, and make ready!" he shouted, and ere the first column was upon them the British infantry had dealt a deadly fire into its ranks which made it pause. The second column was caught in flank by Adam's Brigade. Then two brigades of British cavalry charged, and although the famous Imperial Guards endeavored to hold their own, they were forced back. Blücher, who had arrived at a most opportune moment, carried the position occupied by the French right at Papelotte and La Haye with Ziethen's corps. The whole Allied line then advanced, the heights were carried, and Napoleon's last army, on the victory of which he had staked his all, was scattered. The battle of Waterloo was won. "My plan," said the Commander-in-Chief,

Duke of Wellington

"was to keep my ground until the Prussians appeared, and then to attack the French position; and I executed my plan."

Throughout the night, the Prussians followed up the defeated legions, which got across the Sambre on the 19th. Some 30,000 of Napoleon's men were killed or wounded, and 13,000 of the Allied army, more than half of whom were British. On the 22nd June 1815, the fallen Emperor abdicated in favour of his son; on the 7th July the Allies entered Paris in triumph, and eight days later Napoleon surrendered to Captain Maitland, of H.M.S. *Bellerophon*.

The Desolator desolate! The Victor overthrown! The Arbiter of others fate A Suppliant for his own!

Byron.

CHAPTER XX
Wellington the Statesman (1815–52)

"It is the duty of all to look our difficulties in the face and to lay the ground for getting the better of them."

Wellington.

Wellington.

Outline of specific point of views or inside knowledge in this chapter:

102. See the "Story of Napoleon," p. 135.
103. Not at Wimbledon, as Mr. Asquith said in a speech at the Guildhall in 1911.
104. See Foreword.
105. The point is somewhat obscure owing to conflicting evidence. —See "The Boyhood of a Great King," by A. M. Broadley, pp. 99–100.

While the good folk of London were listening to the guns of the Tower and of the Park, which told of the Waterloo victory, and the joyful news was percolating to the smallest hamlet, Wellington was fighting a battle in which neither sword nor gun was involved. It was one of diplomacy, and he proved the conqueror. Blücher, armed with the Declaration of the 13th March 1815 to the effect that "Napoleon Bonaparte is put beyond the pale of social and civil relations. As enemy and disturber of the repose of the world, he is delivered over to public vengeance," was for seizing the fallen Emperor and shooting him as an outlaw at Vincennes, the scene of the Duc d'Enghien tragedy. **102** The bloodthirsty Prussian asked for Wellington's views on the matter. He received them without delay and expressed in such a way that Blücher must have felt thoroughly ashamed of himself, if only for a passing moment. "They had both acted too distinguished a part in the recent transactions to become executioners," the Duke wrote. This single sentence reveals the sterling uprightness of the man and his hatred of unnecessary bloodshed. Even supposing that he had received authority for the carrying out of such a measure, it is extremely doubtful whether Wellington would have concurred in his colleague's wish. Blücher sneered—and accepted the decision. Wellington also found himself in disagreement with the Prussian view regarding the bridge of Jena at Paris, which commemorated the crushing Prussian defeat of October 1806. Blücher, with true patriotic zeal as it seemed to him, was for blowing it to pieces. Wellington regarded the idea as foolish, and he carried his point. It would have been a

Duke of Wellington

bitter day for the Capital had Blücher been allowed to work his will. The Prussian Commander insisted on levying a contribution on the city of Paris of 100,000,000 francs. Wellington upset the scheme by insisting that the question was one for the Allied sovereigns to arrange. For the third time, vindictive Blücher had to give in.

When the Provisional Government appealed to the Duke concerning Napoleon's successor, he bluntly told them that "the best security for Europe was the restoration of the King," namely, Louis XVIII. He should be recalled "without loss of time, so as to avoid the appearance of the measure having been forced upon them by the allies."

When the Exile King returned to Paris the enormous demands of certain of the Powers, particularly of Austria and Prussia, had to be discussed. Had it not been for the resistance of Castlereagh, Wellington, and Nesselrode, extended partitions undoubtedly would have resulted. As finally settled by the Second Treaty of Paris, concluded on the 20th November, the territory of France was reduced to practically the limits of 1790. What was an indemnity of 700,000,000 francs and was determined for the expenses of the war. An army of occupation not exceeding 150,000 troops under the Duke was to garrison the chief frontier fortresses, including Valenciennes, Cambray, Quesnoy, Maubeuge, and

Landrecy, for a maximum period of five years, the expense being met by the French Government. The magnificent art treasures, which Napoleon had gloried in plundering, were to be returned to their rightful owners. That is why the celebrated bronze horses of St Mark may be seen on the great Venetian Cathedral to-day, and the magnificent "Descent from the Cross," by Rubens, admired in the Cathedral of Antwerp, from whence it had been taken to find a temporary resting-place in the Louvre.

An excellent account of the Army of Occupation is given by a Scotsman, who visited Paris shortly after the battle of Waterloo. There were comparatively few Austrians, the majority of them being in the south of France, but those seen by the writer of *Paul's Letters to his Kinsfolk* were "bulky men" who "want the hardy and athletic look of the British, Russians, or Prussians." The Russian infantry were "adequate, tough, steady-looking men, clean, handsome, but by no means remarkable for stature." The artillery was "of the highest possible order," the cavalry "remarkably fine men," the appearance of the Cossacks "prepossessing." The Prussians, while never having been accused of "gross violence," succeeded in wrecking the Château de Montmorency, where a large body of them was quartered. Camp-kettles were boiled with picture-frames, and the furniture stripped by female camp followers. Paul notes that the Prussian officers were the

Duke of Wellington

principal customers of the expensive restaurants and theatres, but that many British officers of rank had gone so far as to decline the quarters appointed them in private houses. He bestows much praise on Wellington for his discipline and justice: "The strong sense and firmness for which the Duke is as much distinguished as for skill in arms and bravery in the field of battle. He easily saw that the high and paramount part which Britain now holds in Europe, that preeminence which, in so many instances, has made her and her delegates the chosen mediators when disputes occurred amongst the allied powers, depends entirely on our maintaining pure and sacred the national character for good faith and disinterested honour. The slightest complaint, therefore, of want of discipline or oppression perpetrated by a British officer or soldier has instantly met with reprehension and punishment, and the result has been the reducing the French to the hard situation of hating us without having any complaint to justify themselves for doing so, even in their own eyes.... The soldiers, without exception, both British and foreigners, conduct themselves in public with civility, are very rarely to be seen intoxicated, though the means are so much within reach; and, considering all the irritating circumstances that exist, few quarrels occur betwixt them and the populace. Unyielding precautions are, however, taken in case of any accidental or premeditated commotion."

Duke of Wellington

Wellington threw the whole weight of his influence on the side of moderation. Prussia was all for partition, for getting her territorial "pound of flesh," but the calmer statesmanship of the diplomatists already mentioned, especially of Wellington, won the day. The Duke's policy is clearly outlined in his dispatch of the 11th August, which, in the opinion of Dr Holland Rose, "deserves to rank among his highest titles to fame."

Wellington states that while France has been left "in too great strength for the rest of Europe, weakened as all the powers of Europe have been by the wars in which they have been engaged with France." His objection to the demand of a "great cession from France upon this occasion is that it will defeat the object which the Allies have held out to themselves in the present and the preceding wars." He then proceeds to detail what were, in his opinion, the various causes which led to so much bloodshed. "To put an end to the French Revolution, to obtain peace for themselves and their people, to have the power of reducing their overgrown military establishments, and the leisure to attend to the internal concerns of their several nations, and to improve the situation of their people. The Allies took up arms against Buonaparte because it was certain that the world could not be at peace as long as he should possess, or should be in a situation to attain, supreme power in France. Care must be taken," he adds, "in

Duke of Wellington

making the arrangements consequent upon our success, that we do not leave the world in the same unfortunate situation respecting France that it would have been in if Buonaparte had continued in possession of his power."

The Duke then goes on to review the situation. If Louis XVIII were to refuse the cession of territory, his people would undoubtedly support him, and the Allies. "Might take the fortresses and provinces which might suit them, but there would be no genuine peace for the world, no nation could disarm, no Sovereign could turn his attention from the affairs of this country." If the King consented to the partition, "which, from all that one hears, is an event by no means probable, the Allies must be satisfied, and must retire. Nevertheless, I would appeal to the experience of the transactions of last year for a statement of the situation in which we should find ourselves." France was then reduced to her limits of 1792, and "the Allies were obliged to maintain each in the field half of the war establishment stipulated in the treaty of Chaumont, to guard their conquests, and what had been ceded to them." In France, "the general topic of conversation was the recovery of the left bank of the Rhine as the frontier of France." Wellington, therefore, preferred "the temporary occupation of some of the strong places, and to maintain for a time a strong force in France, both at the expense of the French

Duke of Wellington

Government, and under strict regulation. The permanent cession of even all the places which in my opinion ought to be occupied for a time. These measures will not only give us, during the period of occupation, all the military security which could be expected from the permanent session, but, if carried into execution in the spirit in which they are conceived, they are in themselves the bond of peace."

During the remainder of his stay in France, broken by a short visit to England in 1816, Wellington was far from popular, and one or two attempts were made on his life. It was scarce to be expected that a man who had been pre-eminently successful in the field against the nation's armies would be lauded as a widespread hero, but, as we have seen, he had helped to save the country from a bitter and vindictive humiliation. He finally returned home in 1818, when the Army of Occupation evacuated France with the consent of the Powers at the request of its Commander-in-Chief. He never again drew his sword in warfare, but he maintained a commanding position in the affairs of Great Britain until the close of his long life.

His honors and orders were now varied and many. England and foreign countries honored themselves by honoring him. Parliament voted him £200,000 for the erection or purchase of Strathfieldsaye, Hampshire, and the estate granted

Duke of Wellington

to him as Prince of Waterloo by the King of the Netherlands was valued at £4000 per annum. On his return to England, he became Master of the Ordnance, which entitled him to a seat in the Cabinet.

In 1821, Wellington revisited the scene of his most famous battle with George IV and afterwards proceeded to Verona to represent, with Lord Strangford, Great Britain at the Congress. What was held there to determine the attitude of England, France, Russia, Austria, and Prussia regarding various matters, including the insurrection in Greece and the relations between Russia and Turkey. In the matter, the evacuation of Piedmont and Naples by the Austrian troops, the slave trade, and more particularly the unhappy state of affairs in Spain, which country was then in a state of civil war. Should the five Powers send armed assistance to Ferdinand, whom Wellington's victories in the Peninsula had replaced on the throne? Answering for his own country the Duke maintained the principle of non-interference excepting in a case of necessity. In this matter, Great Britain stood alone, and the Duke had to run the gauntlet of fierce criticism on his return to England.

His next continental journey was in 1826 when he was sent on a special mission to Petersburg on the accession of Emperor Nicholas, with the object of arriving at a satisfactory settlement of the projected Russian attack on Turkey over the Greek difficulty. In this, he was not entirely successful, for after events proved that he had only succeeded in staving off the evil day.

On the death of the Duke of York in the following year, Wellington was appointed Commander-in-Chief, retaining his other office, which controlled the artillery and engineers merely.

A month later Canning became Prime Minister, and the Duke was asked to continue as a member of the Cabinet. This request he not only declined, but surrendered his two important offices as well. Mutual suspicion seems to have been the cause of this unexpected event, certainly not jealousy, for Wellington said that he should be "worse than mad if he had ever thought about it for a moment," the "it" referring to his possible appointment as First Lord of the Treasury. Canning did not live long to enjoy the sweets of office, for he died on the following August, and was succeeded by "Prosperity" Robinson, otherwise Lord Goderich, who resigned at the beginning of 1828.

The Duke, once again Commander-in-Chief, was sent for by George IV, and requested to form a Ministry. He obeyed with the instinct of a soldier when ordered by his superior officer, rather than as a keen politician about to have his

Duke of Wellington

highest ambition gratified. Wellington was a Tory, and the political freedom of the Roman Catholics and the reform of Parliament were the burning questions of the hour. The Duke was uncertain as to the practical utility of either, but he was not prepared to go against the public wishes of the nation so far as the religious question was concerned. After navigating a sea of difficulties, the Roman Catholic Relief Bill passed both Houses in the early days of 1829. One of his opponents, the Earl of Winchilsea, charged Wellington with "breaking in upon the Constitution of 1688 in order that he might the more successfully, under the cloak of some outward show of zeal for the Protestant religion, carry on his insidious designs for the infringement of our liberties, and the introduction of Popery in every department of the State." The Premier requested an apology, which was not forthcoming, after that the former demanded "satisfaction," in other words, a duel. Sir Henry Hardinge for Wellington and Lord Falmouth for Winchilsea were the respective seconds.

The meeting took place at Battersea Fields. "Now then, Hardinge," said the Duke, "look sharp and step out the ground. I have no time to waste. Don't stick him up so near the ditch. If I hit him, he will tumble in." The signal was given to fire. Noting that his opponent did not level his pistol on the command being given, the Duke purposely fired wide, and an instant afterwards Winchilsea fired in the air. The latter then produced a written sheet which he called an apology, which had to be altered before it met with Wellington's approval. "Good morning, my Lord Winchilsea; good morning, my Lord Falmouth," cried the Duke as he saluted with two fingers, and, mounting his horse, cantered off.

The Duke had a most thankless task during his administration, so much so that we find him writing. "If I had known in January 1828, one tithe of what I do now, and of what I discovered one month after I was in office, I should never have been the King's Minister, and so have avoided loads of misery. However, I trust God Almighty will soon determine that I have been sufficiently punished for my sins and will relieve me of the unlucky lot which has befallen me. I believe there never was a man who suffered so much for so little purpose."

He had almost as much trouble with the King as had Pitt with George III, and many of his old supporters were indignant with him over the Relief Bill. Wellington vehemently opposed Parliamentary Reform in the face of public opinion, with the result that his Ministry rode to a fall in November 1830.

Two months before he had taken part in the opening ceremony of the Manchester and Liverpool Railway, the first

Duke of Wellington

line to cater for passenger traffic in the British Empire. He rode in one of the two trains which made the first journey, and the fact that they both went in the same direction was the cause of a tragic accident which deprived one of Wellington's friends of his life. The incident occurred at Parkside, where the engines stopped to obtain a supply of water. While the trains were at a standstill, Mr. Huskisson, formerly President of the Board of Trade, got out of the carriage in which he had been travelling and sought Wellington. A minute or two later the train on the different line started. One of the open doors knocked him down, and his right leg was crushed by the locomotive. The Duke and several others ran to the injured man's assistance, but his injuries were such that he only survived a few hours.

Wellington was succeeded as First Lord of the Treasury by Earl Grey, whose Government was speedily defeated by the Reform Bill which it introduced being rejected by the Lords. Riots broke out in London and the provinces; William IV "was frightened by the appearance of the people outside of St James's"; the distinguished Dr Arnold wrote that his "sense of the evils of the times, and to what purpose I am bringing up my children, is overwhelmingly bitter." The King implored the Ministers not to hand in their resignation, the House of Commons carried by a massive majority a vote of confidence in the Government, and the nation showed that it bitterly resented the action of the Lords. There was an attempt at compromise, but the concessions were so trivial from Wellington's point of view that he declined to take part in the negotiations. After further angry scenes in the following session, Grey resigned on the 9th May 1832. It was during this trying period of our national history that the window-panes of Apsley House were stoned and the Duke's life was threatened. **104**

Once again the King requested Wellington to form a new administration, and several meetings were held with that idea in view, but to no purpose. He had to confess that the task was without question impossible: "I felt that my duty to the King required that I should make a great sacrifice of opinion to serve him, and to save his Majesty and the country from what I considered a great evil. Others were not of the same opinion. I failed in performing the service which I intended to perform...." Several resident members of Oxford University, including Professor the Rev. John Keble, impressed by the Duke's devotion, raised funds for the purpose of a bust to commemorate his self-denying conduct. This appreciation of approval greatly pleased Wellington, who announced his intention of sitting for Chantrey, the celebrated sculptor, or whoever else the committee might choose,

Duke of Wellington

"with the greatest satisfaction." When Grey resumed office, the Reform Bill was read for a third time and passed. Some peers having declared "that in consequence of the present state of affairs they have come to the resolution of dropping their further opposition to the Reform Bill so that it may pass without delay as nearly as possible in its present shape." Wellington quietly left the House. He was no kindlier disposed towards the Irish Reform Bill and subjected it to a fire of criticism which did not, however, preclude it from passing.

One of the most remarkable events of the Duke's full life occurred in November 1834. When Earl Grey resigned in July 1834, on which occasion his opponent made a graceful speech to the effect that there had been no personal hostility in his opposition, the retiring statesman recommended Lord Melbourne as his successor. This suggestion met with the King's approval, but the reign of the new Administration lasted only until the middle of the following November. His Majesty sent for Wellington at six in the morning. The latter refused to form a Cabinet and recommended Sir Robert Peel, who was then in Rome. The Duke promised to carry on the Government during the interim, with the result that he held the offices of First Lord of the Treasury, Home Secretary, Foreign and Colonial Secretary, and Secretary at War for nearly a month. On Peel's return, he appointed his industrious ally

Foreign Secretary, a position he held until the following April when the Government resigned. In 1841, in Peel's second Administration, he occupied a seat in the Cabinet, without an office, and in the following year he was created Commander-in-Chief for life by patent under the great seal.

During the Chartist agitation, Wellington was asked who was to command the forces in London, where a riot was expected. He answered, "I can name no one except the Duke of Wellington." He organized the arrangements with his usual thoroughness, disposing his troops to keep them out of sight and taking prompt measures to protect important public buildings. Fortunately, the excitement died down, and armed force was not required.

The Duke frequently spent several hours a day at the Horse Guards. "Speaking from the experience which I had with him," says General Sir George Brown, G.C.B., "I should say that the Duke was a remarkably willing man to do business with, because of his clear and ready decision. However much I may have seen him irritated and excited, with the subjects which I have repeatedly had to bring under his notice. I have no recollection of his ever having made use of a harsh or discourteous expression to me, or of his having dismissed me without a distinct and explicit answer or decision in the case under consideration. Like all good men of business, who consider well before coming to a decision, his Grace was accustomed to adhere strictly to precedent; to the decisions he may have previously come to on similar cases. This practice greatly facilitated the task of those who had to transact business with him, seeing that all we had to do in concluding our statement of any particular case was to refer to his decision or some similar one."

"Everybody writes to me for everything," he once remarked to Stanhope. "They know the Duke of Wellington is said to be a gracious man, and so at the least they will get an answer." The Earl, astonished at the amount of the Duke's correspondence, ventured to say that his host might expect to be allowed some rest and recreation while he was at Walmer. "Rest!" cried the Duke. "Every other animal—even a donkey—a costermonger's donkey—is allowed some rest, but the Duke of Wellington never! There is no help for it. As long as I am able to go on, they will put the saddle on my back and make me go."

Georgiana, Lady De Ros, who was a frequent visitor at Walmer Castle and at Strathfieldsaye, relates an incident which has a direct bearing on this point. "Wellington," she says, "would tell a story against himself sometimes, and amused us all quite in his latter days by the account of various impostures that had been practised upon him. For years, he

Duke of Wellington

had helped an imaginary officer's daughter, paid for music lessons for her, given her a piano, paid for her wedding trousseau, for her child's funeral, etc., etc. At last it came out that *one man* was the author of these impostures, 'and then,' the Duke said, 'an Officer from the Mendicity Society called on me and gave me such a scolding as I never had before in my life!'"

In a book inscribed as "A Slight Souvenir of the Season 1845–6" we find a delightful little glimpse of "the hero of a hundred fights" as a country gentleman. "What can be a finer sight than to see the Duke of Wellington enter the hunting field?" the author asks. "Not one of those gorgeous spectacles, it is true, such as a coronation, a review, the Lord Mayor's Show, or a procession to the Houses of Parliament. However, not one of those pompous Continental exhibitions called a *chasse*, where armed menials keep back the crowd, and brass bands proclaim alike the find and finish. What can be a finer sight—a sight more genial to the mind of a Briton—than the mighty Wellington entering the hunting field with a single attendant, making no more fuss than a country squire? Yet many have seen the sight, and many, we trust, may yet see it. The Duke takes the country sport like a country gentleman—no man less the great man than this greatest of all men; affable to all, his presence adds joy to the scene. The Duke is a true sportsman and has long been a supporter of the Vine and

Duke of Wellington

Sir John Cope's hounds. He kept hounds himself during the Peninsular War, and divers good stories are related of them and their huntsman (Tom Crane), whose enthusiasm used sometimes to carry him into the enemy's country, a fact that he used to be reminded of by a few bullets whizzing about his ears."

Wellington was now the trusted friend of Queen Victoria, whoever held him in the highest esteem. He was one of the first persons, perhaps actually the first, outside the Royal family and the medical attendants to see the baby who afterwards became Edward VII. According to one account he was met outside Buckingham Palace by Lord Hill, who was informed "All over—smart boy, very fine boy, almost as red as you Hill."

Two days after the first anniversary of the birthday of Edward Albert, Prince of Wales, the Queen and the Prince Consort, accompanied by the Royal children, journeyed to Walmer Castle to pay the Duke a visit. An even greater honor was reserved for the veteran warrior, for on the birth of her Majesty's third son on the 1st May 1850, it got noised abroad that the infant was to be called Arthur, "in compliment to the Hero of Waterloo." The present Duke of Connaught is thus a living link with Wellington. "I must not omit to mention," the Queen writes exactly a year later, "a fascinating episode of this day, viz., the visit of the good old Duke on this his eighty-second birthday, to his little godson, our dear little boy. He came to us both at five, and gave him a golden cup and some toys, which he had himself chosen, and Arthur gave him a nosegay."

The day was also that on which the great Exhibition at the Crystal Palace was opened. "The Royal party," says Queen Victoria, "were received with continued acclamation as they passed through the Park and round the Exhibition House, and it was also fascinating to witness the cordial greeting given to the Duke of Wellington. I was just behind him and Anglesey [on whose arm he was leaning], during the procession around the building, and he was accompanied by an incessant running fire of applause from the men. There were waving of handkerchiefs and kissing of hands from the women, who lined the pathway of the march during the three-quarters of an hour that it took us to march round...."

Although the Duke never courted popularity, seemed indeed to shun it and to regard the satisfaction shown by some of his colleagues in the plaudits of the multitude as a sign of weakness. There can be little doubt that he felt a glow of inward pleasure, however slight, when he reflected on the good feeling displayed towards him in the closing years of his

Duke of Wellington

long and well-filled life. Apt to be somewhat cynical on occasion, and to think that the times were "like sweet bells jangled, out of tune and harsh," he was neither censorious nor vindictive. Nelson preached the gospel of Duty, but Wellington lived it and sacrificed everything to it.

Brougham, as a champion of Parliamentary Reform, was an opponent of Wellington, but in middle age he took up an independent position and has left in his "Historical Sketches of Statesmen who flourished in the Time of George III" a magnificent testimony of the Duke's worth.

"The peculiar characteristic of this great man," he writes, "and which, though far less dazzling than his exalted genius. His marvellous fortune is incomparably more useful for the contemplation of the statesman. He is as well as the moralist, is that constant abnegation of all selfish feelings, that habitual sacrifice of every personal, every party consideration to the single object of strict duty—duty rigorously performed in what station soever he might be called to act. This was ever perceived to be his different quality, and it was displayed at every period of his public life, and in all matters from the most trifling to the most important."

Regarding the Reform Bill, Brougham says that Wellington's conduct "during the whole of the debates in both sessions upon that measure was exemplary. Opposing it to the utmost of his power, no one could charge him with making the least approach to factious violence, or with ever taking an unfair advantage.... After the Bill had passed, the same absence of all factious feelings marked his conduct."

The Duke's modesty, his good sense, candor and fairness, love of justice, hatred of oppression and fraud are touched upon by Brougham, who closes his brief acknowledgment of his subject's virtues by quoting a remark made by Lord Denman, "the greatest judge of the day." It is that of all Wellington's "great and good qualities, the one who stands first, is his anxious desire ever to see justice done, and the pain he manifestly feels at the sight of injustice."

On the morning of the 14th September 1852, the victor of Waterloo had a paralytic stroke at Walmer Castle. At six o'clock his valet entered the Duke's room to call him, but he complained of not feeling quite well and sent for an apothecary. In the evening, he was lying dead on his camp bedstead. We are apt to use the phrase "full of years and honor" rather too glibly perhaps, but it is intensely apposite when applied to the great Duke. He was eighty-three years of age, and as for honor a glance at the following list of distinctions bestowed upon Arthur Wellesley will make the fact self-evident:

Duke of Wellington

He was Duke of Wellington, Marquis of Wellington, Earl of Wellington in Somerset, Viscount Wellington of Talavera, Marquis of Douro, Baron Douro of Wellesley, Prince of Waterloo in the Netherlands, Duke of Ciudad Rodrigo in Spain, Duke of Bennoy in France, Duke of Vittoria, Marquis of Torres Vedras, Count of Vimiero in Portugal, a Grandee of the First Class in Spain, a Privy Councillor, Commander-in-Chief of the British Army, Colonel of the Grenadier Guards, Colonel of the Rifle Brigade, a Field Marshal of Great Britain, a Marshal of Russia, Austria, Prussia, Spain, Portugal, and the Netherlands; a Knight of the Garter, the Holy Ghost, the Golden Fleece; a Knight Grand Cross of the Bath and of Hanover, a Knight of the Black Eagle, the Tower and Sword, St Fernando, of William of the Low Countries, Charles III, of the Sword of Sweden, St Andrew of Russia, the Annunciado of Sardinia, the Elephant of Denmark, of Maria Theresa, of St George of Russia, of the Crown of Rue of Saxony; a Knight of Fidelity of Baden, of Maximilian Joseph of Bavaria, of St Alexander Newsky of Russia, of St Hermenegilda of Spain, of the Red Eagle of Bradenburg, of St Januarius, of the Golden Lion of Hesse Cassel, of the Lion of Baden; and a Knight of Merit of Würtemburg. In addition, Wellington was Lord High Constable of England, Constable of the Tower and of Dover Castle, Warden, Chancellor and Admiral of the Cinque Ports, Lord-Lieutenant of Hampshire and of the Tower Hamlets,

Duke of Wellington

Ranger of St James's Park and of Hyde Park, Chancellor of the University of Oxford, Commissioner of the Royal Military College, Vice-President of the Scottish Naval and Military Academy, the Master of Trinity House, a Governor of King's College, a Doctor of Laws, and a Fellow of the Royal Society.

The motto on Wellington's escutcheon,

Virtutis fortuna comes

"Fortune is the companion of valor"—was exemplified in his long and eventful career, and perhaps the following words, once used by him in a dispatch, suggest how keen was his sense of responsibility: "God help me if I fail, for no one else will." With true British inconsistency the nation spent £100,000 on the funeral of him, whose habits were of Spartan simplicity, but with more appropriateness the body of the Conqueror of Napoleon was placed next to that of the Hero of Trafalgar in the crypt of St Paul's Cathedral.

And so these two great Warriors sleep together. They were worthy of England; may England be worthy of them.

Appendix

WATERLOO

This dispatch was written after the battle of Waterloo.

To Marshal Lord Beresford, G. C. B.:

You will have heard of our battle of the 18th. Never did I see such a pounding match. Both were what the boxers call "gluttons." Napoleon did not manoeuver at all. He just moved forward in the old style in columns, and was driven off in the old style. The only difference was, that he mixed cavalry with his infantry, and supported both with an enormous quantity of artillery.

I had the infantry for some time in squares, and I had the French cavalry walking about as if they had been our own. I never saw the British infantry behave so well.

Wellesley.

Duke of Wellington

OPPOSITION TO REFORM

The Prime Minister, the Duke of Wellington, made the attached speech to address the House of Lords in 1830. The speech positioned him as the opposition to reform and in the end defeated the ministry.

This subject brings me to what noble lords have said respecting the putting the country in a state to overcome the evils likely to result from the late disturbances in France. The noble Earl has alluded to the propriety of effecting parliamentary reform. The noble Earl has, however, been candid enough to acknowledge that he is not prepared with any measure of reform, and I can have no scruple in saying that his Majesty's government is as totally unprepared with any plan as the noble Lord. Nay, I, on my own part, will go further, and say, that I have never read or heard of any measure up to the present moment which can in any degree satisfy my mind that the state of the representation can be improved, or be rendered more satisfactory to the country at large than at the present moment. I will not, however, at such an unseasonable time, enter upon the subject, or excite discussion, but I shall not hesitate to declare unequivocally what are my sentiments upon it. I am fully convinced that the country possesses at the present moment a legislature which answers all the good purpose of legislation, and this to a greater degree than any legislature ever has answered in any country whatever. I will go further, and say, that the legislature and the system of representation possess the full and entire confidence of the country—deservedly possess that confidence—and the discussions in the legislature have a very great influence over the opinions of the country. I will go still further, and say, that if at the present moment I had imposed upon me the duty of forming a legislature for any country, and particularly for a country like this, in possession of great property of various descriptions, I do not mean to assert that I could form such a legislature as we possess now, for the nature of man is incapable of

Duke of Wellington

reaching such excellence at once; but my great endeavor would be to form some description of legislature which would produce the same results. The representation of the people at present contains a large body of the property of the country, and in which the landed interests have a preponderating influence. Under these circumstances, I am not prepared to bring forward any measure of the description alluded to by the noble Lord. I am not only not prepared to bring forward any measure of this nature, but I will at once declare that as far as I am concerned, as long as I hold any station in the government of the country, I shall always feel it my duty to resist such measures when proposed by others.

Duke of Wellington

Two-Page Illustrations

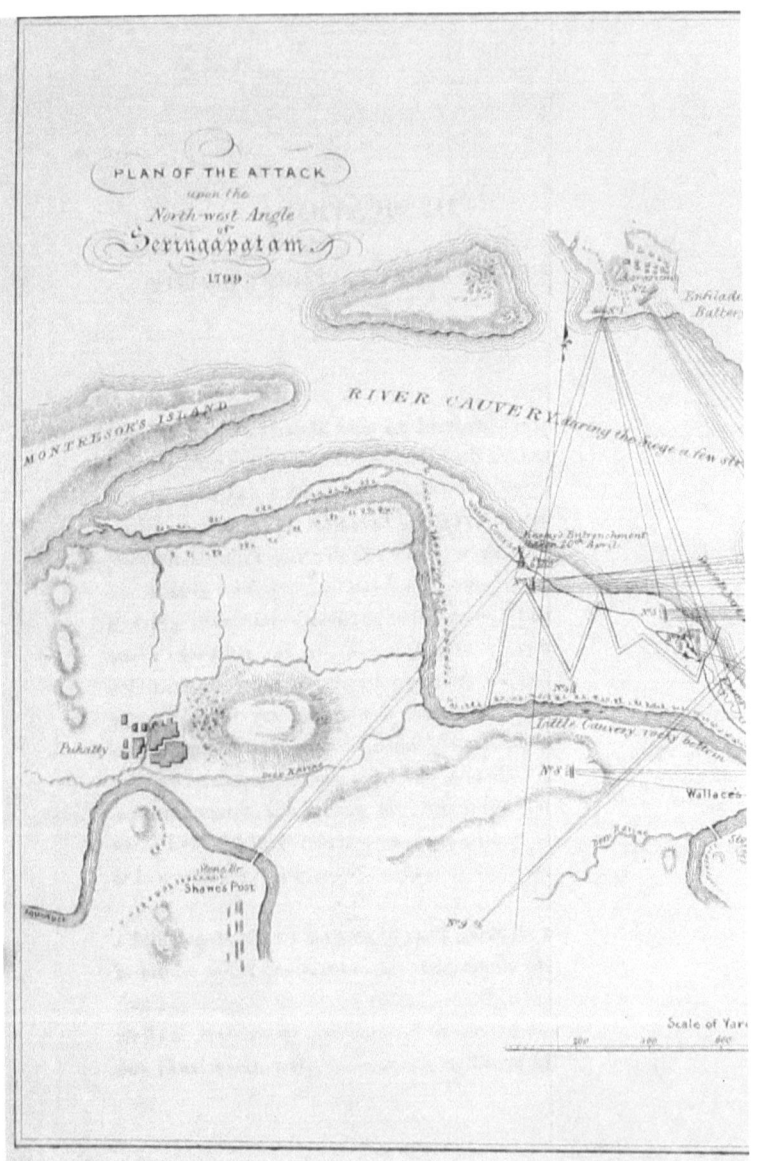

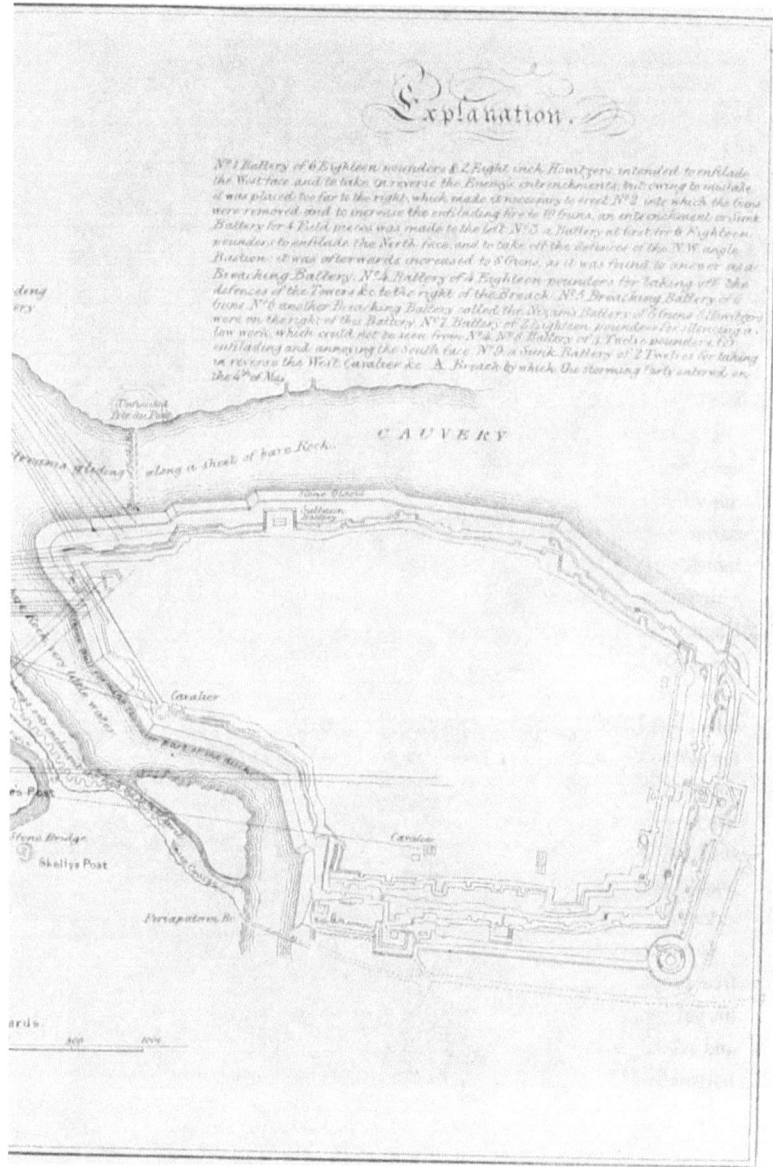

Duke of Wellington

Duke of Wellington

Duke of Wellington

Duke of Wellington

E FLEURUS
iin 1794

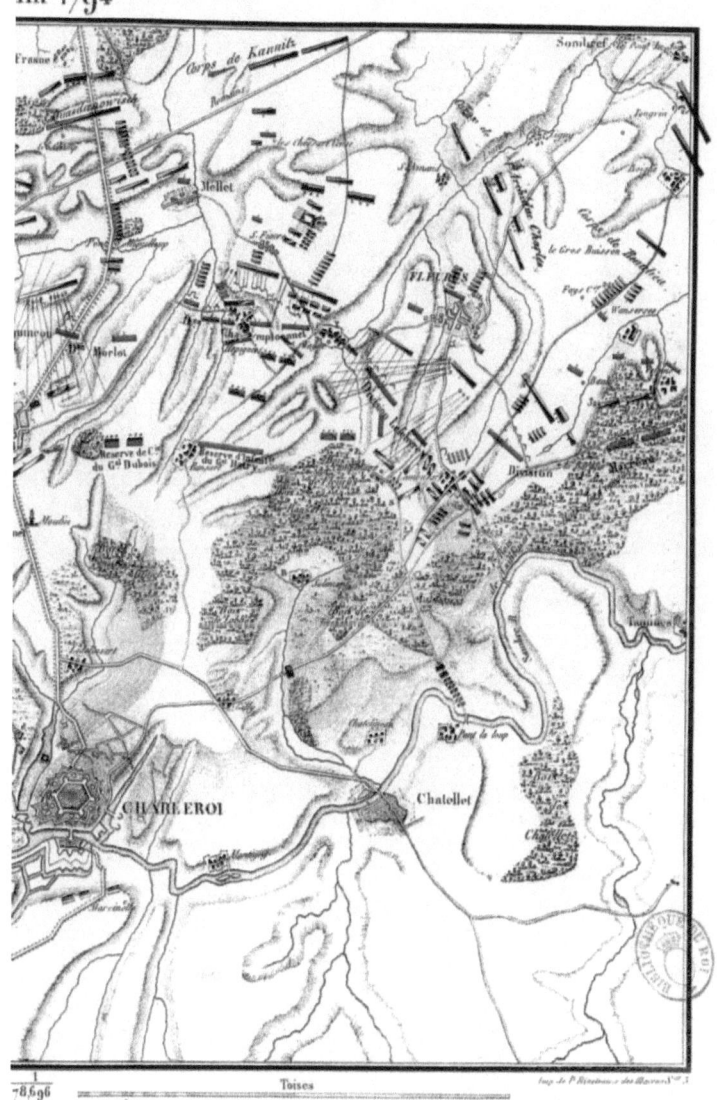

Collection Bibliothèque Nationale, Paris (cartes et plans).

Duke of Wellington

Duke of Wellington

Notes

Duke of Wellington, Southampton

If you visit Southampton and have the opportunity go to the Duke of Wellington, the home-cooked food is very tasty. The staff is very friendly, cozy atmosphere. A typical English pub, well worth the visit.

Duke of Wellington, Notting Hill, London

Go shopping and step into the Duke of Wellington. The food is great and the atmosphere outstanding. You have to try the signature dish Lamb Wellington, however keep in mind it is for two to share. Close to Ladbroke Grove.

Apsley House, London

Duke of Wellington

Is the townhouse of the Duke of Wellington also known as **Number One, London.** Look up and pay a visit, it is located on the south east corner of the Hyde Park.

Duke of Wellington

Figure 28 The French Empire and sphere of influence 1812

Duke of Wellington

bookforces
Publishing - Miami USA

Chris Pope about the Book

"The Duke of Wellington and Napoleon were rivals in life and in death. The two were the supreme commanders of their age; how much do you know about the influence that the Duke of Wellington had on the path of history?"

The book is combined work derived from significant sources joined by twenty-nine illustrations that are unique to the time period. Would the Duke of Wellington have become the enigmatic figure if Waterloo would not have happened?

It seems like that in 1814, after twenty-five years of war in Europe, it was finally coming to an end. The surrender of Emperor Napoleon ended up with his banishment to the small island of Elba.

Soon after. The European supremacies started to reinstate their countries totality and Europe progressed to normality and peace.

However, Napoleon escaped from Elba in March 1815 and landed in France. Only three weeks later he arrived triumphant in Paris and reinstated his title as Emperor of France. His army reunited with him as if the war never happened. The soldiers who had been captured during the time of war were released by the kind nature of the European forces, what enabled Napoleon to reorganization his Grande Armée.

Duke of Wellington

The Duke of Wellington and the European allies had to reunite their armies and arrange to recommence the war to defeat the Emperor yet once again.

Napoleon and Duke of Wellington, whose soldiers fought against one each other meet for their ending battle at Waterloo.

www.ingramcontent.com/pod-product-compliance
Lightning Source LLC
Chambersburg PA
CBHW032147010526
44111CB00035B/1236